Springer Proceedings in Mathematics & Statistics

Volume 243

Springer Proceedings in Mathematics & Statistics

This book series features volumes composed of selected contributions from workshops and conferences in all areas of current research in mathematics and statistics, including operation research and optimization. In addition to an overall evaluation of the interest, scientific quality, and timeliness of each proposal at the hands of the publisher, individual contributions are all refereed to the high quality standards of leading journals in the field. Thus, this series provides the research community with well-edited, authoritative reports on developments in the most exciting areas of mathematical and statistical research today.

More information about this series at http://www.springer.com/series/10533

L. Beilina · Yu. G. Smirnov
Editors

Nonlinear and Inverse Problems in Electromagnetics

PIERS 2017, St. Petersburg, Russia, May 22–25

Springer

Editors
L. Beilina
Department of Mathematical Sciences
Chalmers University of Technology
and University of Gothenburg
Gothenburg, Sweden

Yu. G. Smirnov
Penza State University
Penza, Russia

ISSN 2194-1009 ISSN 2194-1017 (electronic)
Springer Proceedings in Mathematics & Statistics
ISBN 978-3-319-94059-5 ISBN 978-3-319-94060-1 (eBook)
https://doi.org/10.1007/978-3-319-94060-1

Library of Congress Control Number: 2018945433

Mathematics Subject Classification (2010): 65M06, 65M12, 65M15, 65M32, 65M50, 65M55, 65M60, 65M70, 65M80

Printed on acid-free paper

This Springer imprint is published by the registered company Springer Nature Switzerland AG
The registered company address is: Gewerbestrasse 11, 6330 Cham, Switzerland

Preface

In this volume are collected some of the chapters related to the special session *Nonlinear and Inverse Problems in Electromagnetics* which was organized at the conference *Progress in Electromagnetics Research Symposium, PIERS 2017* and took place from May 22 to 25, 2017 in St. Petersburg, Russia. The purpose of this session was to provide a forum for academic discussion on the theory and practice of mathematical analysis of nonlinear and inverse problems arising in electromagnetics, and the applications of such problems.

All papers in this volume highlight the most recent research related to the new efficient numerical methods for the solution of nonlinear and inverse problems linked to electromagnetics. The importance of this topic is due to a wide range of applications, such as many aspects of acoustics, optics, medical imaging and reconstruction, geophysics. It is well known that inverse problems are both nonlinear and ill-posed. These two factors cause major challenges for the development of new numerical methods for the solution of nonlinear and inverse problems associated with electromagnetics which are widely presented in the book. Most of chapters of this book provide in a compact form a complete approach for the solution of electromagnetic problems, from correct mathematical problem statement to numerical results. The book contains descriptions of recently developed theoretical and numerical methods for the solution of nonlinear and inverse problems connected with electromagnetics and provides numerical verification of the proposed new numerical methods on simulated and experimental data.

Optimization analysis of a two-dimensional magnetic cloaking problem for bilayer structure is presented in the paper by G. V. Alekseev and Y. E. Spivak. Authors show the solvability of direct and optimization problems for the magnetic scattering model and prove local uniqueness and stability of optimal solutions. Application of the waveguide method in mathematical modeling of multilayered radar absorbing coating is presented in the paper by E. D. Derevyanchuk, A. S. Shutkov, and Yu. G. Smirnov. Authors discuss a numerical–analytical method for the solution of this problem and show numerical results which illustrate their method. The paper by R. O. Evstigneev and M. Yu. Medvedik considers an inverse problem of determination of a complex dielectric permittivity function by measurements of near field outside the body. Authors present the method for the solution of this inverse problem and show numerical results which demonstrate stability of approximate solutions. Diffraction of a plane wave by a dielectric layer filled with a nonlinear medium is presented in the work by Yu. G. Smirnov, E. Smolkin, and V. Kurseeva. Authors develop analytical and numerical solutions of this problem and show numerical results. General nonlinear multi-parameter eigenvalue problems for a nonlinear operator are considered in the work by L. Angermann, Yu. V. Shestopalov, Yu. G. Smirnov, and V. V. Yatsyk. Authors derive functional dispersion equations (solutions of these equations are desired eigenvalues) and prove existence of their roots. Propagation of monochromatic nonlinear symmetric hybrid waves in a cylindrical nonlinear dielectric waveguide is considered in the work by E. Smolkin. The two-step method for reconstruction of an unknown refractive index of an inhomogeneous body is presented in the paper by M. Yu. Medvedik, Yu. G. Smirnov, and A. A. Tsupak. An inverse problem for determination of conductivity function in a hyperbolic equation using single space–time domain noisy observations at the backscattering boundary of the computational domain is considered in the paper by L. Beilina and K. Niinimäki. Authors use Lagrangian approach for the solution of this problem and show numerical results which demonstrate efficiency of the proposed method. The paper by L. Beilina, G. Guillot, and K. Niinimäki uses a finite element method for the problem of magnetic resonance imaging (MRI). Authors formulate a finite element method to solve this problem, present an adaptive algorithm for local mesh refinements, and show reconstruction results from experimental MR data measured by 4.7 Tesla scanner at the laboratory IR4M, CNRS, Université Paris-Sud, France. The work by L. Beilina, M. Cristofol, and S. Li is devoted to the reconstruction of the time- and space-dependent coefficient in an inverse hyperbolic problem in a bounded domain. Authors prove the uniqueness and a Hölder stability in determining of the conductivity function using a single measurement on the lateral boundary and show results of numerical reconstruction.

We can note that as one of the book features is that in chapters a reader finds the whole description of used numerical method, from problem statement to numerical results. The intended audience of the book are: university students (knowledge of mathematics: bachelor level and higher), Ph.D. students (specializing in applied mathematics, mathematics, electrical engineering, physics), Dr. Sci., researchers, university teachers, R&D engineers, and electrical engineers with deeper knowledge and interest in mathematics.

Gothenburg, Sweden L. Beilina
Penza, Russia Yu. G. Smirnov

Contents

Optimization Analysis of a 2D Magnetic Cloaking Problem for Bilayer Structure

G. V. Alekseev and Y. E. Spivak

Abstract We consider the control problems for the 2D model of magnetic scattering by a permeable isotropic obstacle having the form of a cylindrical bilayer. These problems arise while developing the design technologies of magnetic cloaking bilayer devices using the optimization method for solving the corresponding inverse problems. The solvability of direct and optimization problems for the magnetic scattering model under study is proved. The optimality system which describes the necessary conditions of extremum is derived. Based on its analysis the sufficient conditions to the data are established which provide local uniqueness and stability of optimal solutions. Also numerical aspects of applying the optimization approach are discussed.

Keywords Magnetic scattering model · Cloaking problem · Optimization method · Control problem · Solvability · Stability estimates

1 Introduction

In recent years significant research has focused on design of invisibility cloaking devices for material bodies. Beginning with pioneering papers [1–4] the large number of publications was devoted to developing different methods of solving the cloaking problems. The first works in this field focused on the electromagnetic cloaking, i.e. cloaking objects from detection by electromagnetic location. Then the main results of the electromagnetic cloaking theory were expanded to an acoustic cloaking [5, 6] and to cloaking magnetic, electric, thermal and other static fields [7–13]. It should

G. V. Alekseev (✉) · Y. E. Spivak
Institute of Applied Mathematics Far Eastern Branch of RAS,
7, Radio Street, Vladivostok, Russia
e-mail: alekseev@iam.dvo.ru

G. V. Alekseev · Y. E. Spivak
Far Eastern Federal University, 8, Sukhanova Street, Vladivostok, Russia
e-mail: uliyaspivak@gmail.com

L. Beilina and Yu. G. Smirnov (eds.), *Nonlinear and Inverse Problems in Electromagnetics*, Springer Proceedings in Mathematics & Statistics 243,
https://doi.org/10.1007/978-3-319-94060-1_1

be emphasized that the technical realization of solutions obtained in these papers is connected with substantial difficulties.

One of approaches of overcoming these difficulties consists of replacing "exact" singular solutions of cloaking problems under study by approximate non-singular solutions and designing cloaking devices based on these approximations (see e.g., [14–17]). Another approach is based on using the optimization method of solving inverse problems. This approach is based on replacing initial inverse cloaking problem by minimization problem of a suitable tracking-type cost functional which corresponds to inverse problem under consideration. This method to which one should referred to as inverse design method [18, 19] was applied in [20–22] devoted to the numerical analysis of 2D problems of designing layered cloaking shells. Besides, it was applied in [19, 23–29] while theoretical study of electromagnetic cloaking problems. Papers [30, 31] are devoted to studying invisibility problem in X-ray tomography.

The goal of this paper is theoretical analysis of the cloaking problem for the 2D model of magnetic scattering by a bilayer shell using the optimization method. Keeping this goal in the mind the plan of our paper is as follow. Firstly, the unique solvability of the direct problem of magnetic scattering by a permeable isotropic obstacle having the form of a cylindrical bilayer with variable permeability $\mu_k, k = 1, 2$ in every layer is proved. Then control problems are formulated. These problems arise while using optimization method for solving corresponding inverse problems for the model under study. The solvability of these control problems is proved, the optimality system is derived which describes the necessary conditions of the extremum. Based on analysis of the optimality system the local uniqueness and stability of optimal solutions are established. Also numerical aspects of applying the optimization approach for solving the magnetic cloaking problems under study are discussed.

2 Statement of Direct Magnetic Scattering Problem

In this paper we consider the problem of designing a magnetic cylindrical bilayer cloaking shell consisting of an infinitely long along the z-axis coaxial cylinder with a cross-section in the form of a ring $R_0 < r < R_1$, surrounded by another coaxial cylinder with a cross-section in the form of a ring $R_1 < r < R_2$. We define the following domains in the plane R^2 (see Fig. 1):

$$\Omega_0 = \{(x, y) : x^2 + y^2 < R_0^2\},\ \Omega_1 = \{(x, y) : R_0^2 < x^2 + y^2 < R_1^2\},$$

$$\Omega_2 = \{(x, y) : R_1^2 < x^2 + y^2 < R_2^2\},\ \Omega_3^\infty = \{(x, y) : R_2^2 < x^2 + y^2 < \infty\}$$

and assume that each of domains $\Omega_0, \Omega_1, \Omega_2, \Omega_3^\infty$ is filled by isotropic medium with permeability $\mu_l, l = 0, 1, 2, 3$. Moreover, we assume that $\mu_0 = \mu_3 = \text{const} > 0$

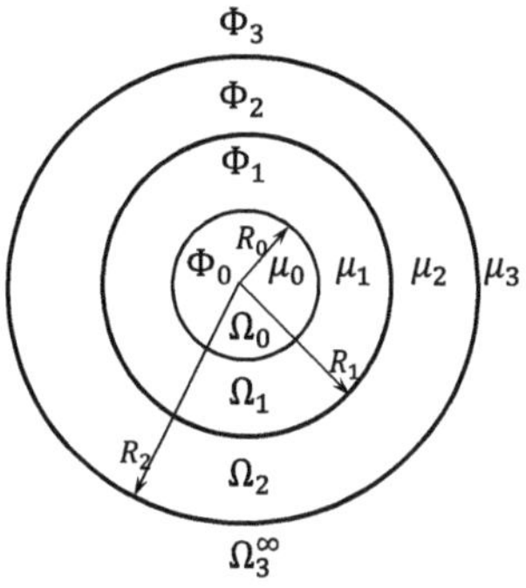

Fig. 1 Schematic diagram of the cross-section of a magnetic bilayer cylinder cloak

while μ_1 and μ_2 are functions of $\mathbf{x} = (x, y)$, describing inhomogeneous isotropic media in Ω_1 and Ω_2, respectively.

Denote by B_R the disk $|\mathbf{x}| < R$ containing domains Ω_0, Ω_1 and Ω_2. We will assume that there are sources outside B_R, which generate externally applied magnetic field $\mathbf{H}_a = -\text{grad}\Phi_a$ corresponding to potential Φ_a (see Fig. 2). Due to the absence of free currents in cloaking shell, magnetic field in the plane R^2 can be expressed by means of a magnetic scalar potential Φ in the form $\mathbf{H} = -\text{grad}\Phi$, where the potential Φ is a solution of equation

$$-\operatorname{div}(\mu\mathbf{H}) = \operatorname{div}(\mu \text{grad}\Phi) = 0. \tag{1}$$

Here μ is a given function, satisfying the conditions $\mu = \mu_0$ in $\Omega_0 \cup \Omega_3^\infty$, $\mu = \mu_1$in Ω_1, $\mu = \mu_2$ in Ω_2. Denote by Φ_l the restriction of potential Φ to the domain Ω_l, $l = 0, 1, 2, 3$. We set $\Phi_s = \Phi_3 - \Phi_a$ and note that the fields Φ_0, Φ_1, Φ_2 and $\Phi_3 = \Phi_a + \Phi_s$ are the solution of the following magnetic scattering problem [19, Chap. 3]:

$$\begin{aligned}\mu_0\Delta\Phi_0 = 0 \text{ in } \Omega_0,\ \operatorname{div}(\mu_1\nabla\Phi_1) &= 0 \text{ in } \Omega_1,\ \operatorname{div}(\mu_2\nabla\Phi_2) = 0 \text{ in } \Omega_2,\\ \mu_3\Delta\Phi_3 &= 0 \text{ in } \Omega_3^\infty,\end{aligned} \tag{2}$$

$$\Phi_0 = \Phi_1,\ \mu_0\frac{\partial\Phi_0}{\partial r} = \mu_1\frac{\partial\Phi_1}{\partial r} \text{ at } r = R_0,\ \Phi_1 = \Phi_2,\ \mu_1\frac{\partial\Phi_1}{\partial r} = \mu_2\frac{\partial\Phi_2}{\partial r} \text{ at } r = R_1,$$

$$\Phi_2 = \Phi_3,\ \mu_2\frac{\partial\Phi_2}{\partial r} = \mu_3\frac{\partial\Phi_3}{\partial r} \text{ at } r = R_2, \tag{3}$$

$$\Phi_s(\mathbf{x}) = O(1) \text{ as } r = |\mathbf{x}| \to \infty. \tag{4}$$

Consider firstly the case when the field $\mathbf{H}_a$ is constant and is directed along the y axis (see Fig. 2) and, besides, $\mu_1 = \text{const}$, $\mu_2 = \text{const}$. Using the Fourier method for finding the solution $(\Phi_0, \Phi_1, \Phi_2, \Phi_3)$ of (2)–(4) one can obtain (see details in [15, 19])

$$\Phi_0(r, \varphi) = \beta_6 r\cos\varphi,\ \ \Phi_1(r, \varphi) = (\beta_4 r + \beta_5/r)\cos\varphi, \tag{5}$$

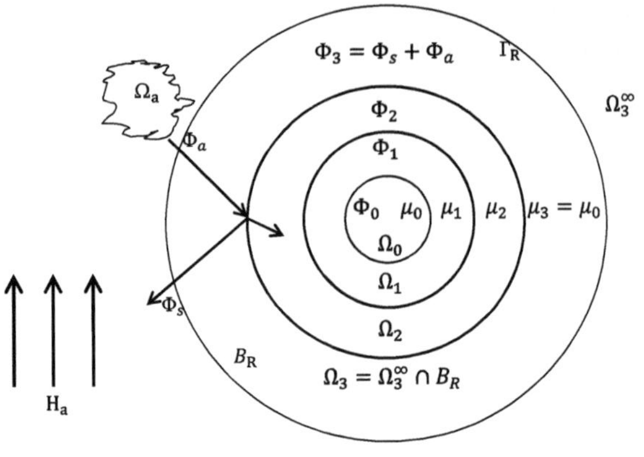

Fig. 2 Schematic illustration of the magnetic shell and the artificial boundary Γ_R in the case of uniform external applied field

$$\Phi_2(r, \varphi) = (\beta_2 r + \beta_3/r)\cos\varphi, \quad \Phi_3(r, \varphi) = (-H_a r + \beta_1/r)\cos\varphi, \quad H_a = |\mathbf{H}_a|. \tag{6}$$

Coefficients β_1, β_2, β_3, β_4, β_5 and β_6 are the solution of linear algebraic system:

$$\begin{gathered}
\beta_2 R_1^2 + \beta_3 - \beta_4 R_1^2 - \beta_5 = 0, \\
\beta_4 R_0^2 + \beta_5 - \beta_6 R_0^2 = 0, \\
-\beta_1 - \mu_2\beta_2 R_2^2 + \mu_2\beta_3 = H_a R_2^2, \\
\beta_1 - \beta_2 R_2^2 - \beta_3 = H_a R_2^2, \\
\mu_2\beta_2 R_1^2 - \mu_2\beta_3 - \mu_1\beta_4 R_1^2 + \mu_1\beta_5 = 0, \\
\mu_1\beta_4 R_0^2 - \mu_1\beta_5 - \beta_6 R_0^2 = 0.
\end{gathered} \tag{7}$$

The determinant of the matrix system (7) is not equal to zero, and hence (7) has a unique solution. Solving the system (7) and substituting the obtained values of coefficients $\beta_1, \ldots, \beta_6$ in (5), (6), we find the required fields Φ_0, Φ_1, Φ_2, Φ_3. With the program Wolfram Mathematica, it is possible to write explicit expressions for all coefficients $\beta_1, \ldots, \beta_6$ in analytical form via parameters μ_1, μ_2, R_0, R_1 and R_2. The main role is played by coefficients β_1 and β_6: in fact, β_6 describes the magnetic field in the internal domain Ω_0 while β_1 describes the perturbation of externally applied magnetic potential $-H_a r \cos\varphi$ in domain Ω_3 equalled to $(\beta_1/r)\cos\varphi$ which is caused by presence of a bilayer cloaking shell $(\Omega_1, \Omega_2, \mu_1, \mu_2)$. We remind that the bilayer shell $(\Omega_1, \Omega_2, \mu_1, \mu_2)$ is a perfect cloak in the case when $\beta_1 = 0$ and $\beta_6 = 0$ [19, p. 91].

In the particular case where the inner layer is a superconductor, so that $\mu_1 = 0$, system (7) is simplified. In particular, β_1 and β_6 are determined as follows [15, 16]

$$\beta_1 = -H_a R_2^2 + \frac{2H_a R_2^2(R_1^2 - R_2^2)\cdot\mu_2}{R_1^2(-1+\mu_2) - R_2^2(1+\mu_2)}, \quad \beta_6 = 0. \tag{8}$$

Furthermore, one easily can show that in the case $\mu_1 = 0$ the condition $\beta_1 = 0$ is carried out if and only if $\mu_2 = \mu_2^0 = (R_2^2 + R_1^2)/(R_2^2 - R_1^2)$. Thus, in the case when $\mu_1 = 0$ and $\mu_2 = \mu_2^0$ the respective shell $(\Omega_1, \Omega_2, \mu_1, \mu_2)$ has exact cloaking effect relatively externally applied field $\mathbf{H}_a$. But this result takes place only in the case when the field $\mathbf{H}_a$ is uniform.

Consider now the case when the external field $\mathbf{H}_a$ is inhomogeneous and therefore it is not possible to apply Fourier's method for finding the exact solution of problem (2)–(4). A number of functional spaces will be used while studying direct problem (2)–(4) and respective control problems. Let $\Omega_3 = \Omega_3^\infty \cap B_R$. We will use the space $H^1(D)$, where D is one of domains B_R, Ω_0, Ω_1, Ω_2, Ω_3, and also spaces $L^\infty(\Omega_k)$, $H^s(\Omega_k)$, $s > 0$, $k = 1, 2$, $L^2(Q)$, $H^{1/2}(\Gamma_R)$ and $H^{-1/2}(\Gamma_R)$. Here $Q \subset B_R$ is an arbitrary open subset of B_R, Γ_R is a boundary of B_R (see Fig. 2). The norms and scalar products in $H^1(D)$, $H^s(\Omega_k)$, $L^2(Q)$, $H^{1/2}(\Gamma_R)$ and $H^{-1/2}(\Gamma_R)$ will be denoted by $\|\cdot\|_{1,D}$, $(\cdot,\cdot)_{1,D}$, $\|\cdot\|_{s,\Omega_k}$, $(\cdot,\cdot)_{s,\Omega_k}$, $\|\cdot\|_Q$, $(\cdot,\cdot)_Q$, $\|\cdot\|_{1/2,\Gamma_R}$ and $\|\cdot\|_{-1/2,\Gamma_R}$. We set $L^\infty_{\mu_0}(\Omega_k) = \{\mu \in L^\infty(\Omega_k) : \mu(\mathbf{x}) \geq \mu_0\}$, $H^s_{\mu_0}(\Omega_k) = \{\mu \in H^s(\Omega_k) : \mu(\mathbf{x}) \geq \mu_0\}$, $k = 1, 2$, $\mu_0 = \text{const} > 0$. It is well known by embedding theorem that the continuous and compact embedding $H^s(\Omega_k) \subset L^\infty(\Omega_k)$ at $s > 1$ holds and the following estimate takes place:

$$\|\mu\|_{L^\infty(\Omega_k)} \leq C_s\|\mu\|_{s,\Omega_k} \quad \forall \mu \in H^s(\Omega_k), s > 1, \ k = 1, 2. \tag{9}$$

C_s is a constant depending on $s > 1$ and Ω_1, Ω_2. We need also a subspace $H(\Omega_3) = \{\Phi \in H^1(\Omega_3) : \Delta\Phi = 0 \text{ in } \Omega_3\}$, equipped with the norm $\|\cdot\|_{\Omega_3} = \|\cdot\|_{H^1(\Omega_3)}$. The space $H(\Omega_3)$ will be serve for describing restrictions of externally applied fields Φ_a to the domain Ω_3.

It should be noted that by trace theorem there exists a trace $\Phi|_{\Gamma_R} \in H^{1/2}(\Gamma_R)$ for any function $\Phi \in H^1(B_R)$ while for any function $\Phi^e \in H(\Omega_3)$ there exists a normal trace $\partial\Phi^e/\partial n|_{\Gamma_R} \in H^{-1/2}(\Gamma_R)$ and the following estimates hold:

$$\|\Phi\|_{1/2,\Gamma_R} \leq C_R\|\Phi\|_X \quad \forall \Phi \in H^1(B_R), \tag{10}$$

$$\|\partial\Phi^e/\partial n\|_{-1/2,\Gamma_R} \leq C'_R\|\Phi^e\|_{1,\Omega_3} \quad \forall \Phi^e \in H(\Omega_3). \tag{11}$$

C_R, C'_R are constants dependent on Ω_3 and R but are independent of $\Phi \in X$ and $\Phi^e \in H(\Omega_3)$.

We assume below that the following conditions take place:

(i) $\mu_1 \in L^\infty_{\mu_1^0}(\Omega_1), \mu_2 \in L^\infty_{\mu_2^0}(\Omega_2), \mu_1 \geq \mu_1^0, \mu_2 \geq \mu_2^0, \mu_1^0 = \text{const} > 0, \mu_2^0 = \text{const} > 0$;
(ii) $\Phi^e \equiv \Phi_a|_{\Omega_3} \in H(\Omega_3)$.

As the potential Φ is determined up to an additive constant we will not distinguish functions of the space $H^1(B_R)$ which differ from each other by an additive constant. Thus the main role below will be played by the following quotient-space $X = H^1(B_R) \setminus R$ with the norm:

$$\|\Phi\|_X^2 = \|\nabla\Phi\|_{B_R}^2 \equiv \|\nabla\Phi\|_{\Omega_0}^2 + \|\nabla\Phi\|_{\Omega_1}^2 + \|\nabla\Phi\|_{\Omega_2}^2 + \|\nabla\Phi\|_{\Omega_3}^2. \tag{12}$$

One can show that the space $X = H^1(B_R) \setminus R$ is Hilbert for this norm and, besides, the following analogue of Poincaré-Friedrichs inequality holds:

$$\|\Phi\|_{B_R} \le C_P \|\Phi\|_X \ \ \forall \Phi \in X. \tag{13}$$

Here C_P is a constant independent of $\Phi \in X$.

We begin our analysis with defining weak formulation and weak solution of direct problem (2)–(4). Preliminarily we reduce problem (2)–(4) to an equivalent boundary problem considered in the bounded domain (disk) B_R.To this end we introduce as in [27] the Dirichlet-to-Neumann operator $T : H^{1/2}(\Gamma_R) \to H^{-1/2}(\Gamma_R)$ which maps every function $h \in H^{1/2}(\Gamma_R)$ to the function $\partial\tilde{\Phi}/\partial\nu \in H^{-1/2}(\Gamma_R)$. Here $\tilde{\Phi}$ is a solution of the external Dirichlet problem for equation $\Delta\tilde{\Phi} = 0$ in $\Omega_3^\infty \backslash B_R$ with the boundary condition $\tilde{\Phi}|_{\Gamma_R} = h$ satisfying the condition $\tilde{\Phi}(\mathbf{x}) = O(1)$ at $r = |\mathbf{x}| \to \infty$. We note that problem (2)–(4) considered in R^2 is equivalent to boundary value problem (2), (3) considered in the disc B_R under the following additional condition for Φ_s on Γ_R:

$$\partial\Phi_s/\partial n = T\Phi_s \text{ on } \Gamma_R. \tag{14}$$

We will refer below to the problem (2), (3) and (14) as Problem 1.

Rely on the space X we derive now the weak formulation of Problem 1. Let $S \in X$ be a test function. We multiply every of equations in (2) considered in Ω_l by the restriction $S|_{\Omega_l}$, $l = 0, 1, 2, 3$, integrate over Ω_l and apply Green formulae. Adding the obtained identities and using the boundary conditions in (3) and (14) we arrive at the following identity for the quadruple $\Phi = (\Phi_0, \Phi_1, \Phi_2, \Phi_3) \in X$:

$$a_\mu(\Phi, S) \equiv a_0(\Phi, S) + a(\mu_1, \mu_2; \Phi, S) = \langle F, S\rangle \ \ \forall S \in X. \tag{15}$$

Here μ denotes the pair (μ_1, μ_2), $a_0(\cdot, \cdot)$, $a(\mu_1, \mu_2; \cdot, \cdot)$ and F are bilinear and linear forms defined by

$$a_0(\Phi, S) = \mu_0 \int_{\Omega_0\cup\Omega_3} \nabla\Phi \cdot \nabla S d\mathbf{x} - \int_{\Gamma_R} (T\Phi) S d\sigma, \tag{16}$$

$$a(\mu_1, \mu_2; \Phi, S) = a_1(\mu_1; \Phi, S) + a_2(\mu_2; \Phi, S), \tag{17}$$

$$a_k(\mu_k; \Phi, S) = \int_{\Omega_k} \mu_k \nabla\Phi \cdot \nabla S d\mathbf{x},\ \langle F, S\rangle = -\int_{\Gamma_R} T\Phi^e S d\sigma + \int_{\Gamma_R} (\partial\Phi^e/\partial n) S d\sigma. \tag{18}$$

Identity (15) represents the weak formulation of problem (2), (3) and (14) and its solution $\Phi = (\Phi_0, \Phi_1, \Phi_2, \Phi_3) \in X$ will be called a weak solution of Problem 1. Arguing as in [27], on can easily show, that the introduction of the weak solution is admissible in the following sense: it satisfies all equations in (2) in the distribution sense, and also boundary conditions in (3) and (14) in the trace sense.

Using Hölder inequality and definition (12) for the norm $\|\cdot\|_X$ we have

$$\left|\int_{\Omega_1} \mu_1 \nabla\Phi \cdot \nabla S d\mathbf{x}\right| \leq \|\mu_1\|_{L^\infty(\Omega_1)} \|\nabla\Phi\|_{\Omega_1} \|\nabla S\|_{\Omega_1} \leq \|\mu_1\|_{L^\infty(\Omega_1)} \|\Phi\|_X \|S\|_X,$$

$$\int_{\Omega_2} \mu_2 \nabla\Phi \cdot \nabla S d\mathbf{x} \leq \|\mu_2\|_{L^\infty(\Omega_2)} \|\nabla\Phi\|_{\Omega_2} \|\nabla S\|_{\Omega_2} \leq \|\mu_2\|_{L^\infty(\Omega_2)} \|\Phi\|_X \|S\|_X.$$

It follows from these estimates and (17) that

$$|a(\mu_1, \mu_2; \Phi, S)| \leq (\|\mu_1\|_{L^\infty(\Omega_1)} + \|\mu_2\|_{L^\infty(\Omega_2)}) \|\Phi\|_X \|S\|_X \quad \forall \Phi, S \in X. \tag{19}$$

Besides, using (10), (11), we derive from (16), (17) and (18) that

$$\left|\mu_0 \int_{\Omega_0 \cup \Omega_3} \nabla\Phi \cdot \nabla S d\mathbf{x}\right| \leq \mu_0 \|\Phi\|_X \|S\|_X, \tag{20}$$

$$\left|\int_{\Gamma_R} (T\Phi) S d\sigma\right| \leq \|T\Phi\|_{-1/2,\Gamma_R} \|S\|_{1/2,\Gamma_r} \leq C_T C_R^2 \|\Phi\|_X \|S\|_X, \tag{21}$$

$$|a_0(\Phi, S)| \leq (\mu_0 + C_T C_R^2) \|\Phi\|_X \|S\|_X, \tag{22}$$

$$|\langle F, S\rangle| \leq (\|T\| \|\Phi^e\|_{1/2,\Gamma_R} + \|\partial\Phi^e/\partial n\|_{-1/2,\Gamma_R}) \|S\|_{1/2,\Gamma_R} \leq$$
$$\leq (C_T C_R + C_R') C_R \|\Phi^e\|_{1,\Omega_3} \|S\|_X. \tag{23}$$

From (19)–(23) follows that forms a_0 and F are continuous on X, and the following estimates hold:

$$\|a_0\| \leq C_0, \; \|F\|_{X^*} \leq C_0 \|\Phi^e\|_{1,\Omega_3}, \; C_0 = \max[\mu_0 + C_T C_R^2, (C_T C_R + C_R') C_R]. \tag{24}$$

X^* is a dual of X with respect to space $L_0^2(\Omega_3)$.

Let us assume, moreover, that $\mu_1 \in L^\infty_{\mu_1^0}(\Omega_1)$, $\mu_2 \in L^\infty_{\mu_2^0}(\Omega_2)$ where $\mu_1^0 > 0$, $\mu_2^0 > 0$. Then we have

$$\int_{\Omega_1} \mu_1 \nabla\Phi \cdot \nabla\Phi d\mathbf{x} \geq \mu_1^0 \|\nabla\Phi\|^2_{\Omega_1}, \quad \int_{\Omega_2} \mu_2 \nabla\Phi \cdot \nabla\Phi d\mathbf{x} \geq \mu_2^0 \|\nabla\Phi\|^2_{\Omega_2},$$

$$a_0(\Phi, \Phi) = \mu_0 \int_{\Omega_0 \cup \Omega_3} |\nabla\Phi|^2 d\mathbf{x} - \int_{\Gamma_R} (T\Phi)\Phi d\sigma \geq \mu_0 \int_{\Omega_0 \cup \Omega_3} |\nabla\Phi|^2 d\mathbf{x} \; \forall \Phi \in X.$$

From these estimates, (19), (24) and the definition of the norm (12) in space X that the bilinear form a_μ defined in (15) is coercive on X under condition (i) and, besides,

$$a_\mu(\Phi, \Phi) \geq \mu_* \|\Phi\|_X^2 \; \forall \Phi \in X, \quad \mu_* = \min(\mu_0, \mu_1^0, \mu_2^0). \tag{25}$$

We note that the bilinear form a_μ, defined in (15), defines a linear operator $A_\mu : X \to X^*$ acting by formula

$$\langle A_\mu \Phi, S \rangle = a_\mu(\Phi, S) \equiv a_0(\Phi, S) + a(\mu_1, \mu_2; \Phi, S), \tag{26}$$

while problem (15) is equivalent to operator equation

$$A_\mu \Phi = F. \tag{27}$$

It follows from (19), (22) that under condition (i) bilinear form a_μ is continuous on X and from (25) follows that a_μ is coercive on X. Then from the Lax-Milgram theorem follows that the operator $A_\mu : X \to X^*$ is an isomorphism and the inverse $A_\mu^{-1} : X^* \to X$ to operator A_μ is also isomorphism. Setting $C_\mu = \|A_\mu^{-1}\|$ we note that by the Lax-Milgram theorem we have $C_\mu \le C_1 = (1/\mu_*)$. Using this estimate we derive that for any element $F \in X^*$ operator Eq. (27) has a unique solution $\Phi_\mu = A_\mu^{-1}(F) \in X$, for which the estimate $\|\Phi_\mu\|_X \le C_1 \|F\|_{X^*}$ holds. Using this fact and the second estimate in (24) we conclude that for any field $\Phi^e \in H(\Omega_3)$ problem (15) has a unique solution $\Phi_\mu \in X$, and the following estimate holds:

$$\|\Phi_\mu\|_X \le C_2 \|\Phi^e\|_{1,\Omega_3}, \quad C_2 = C_0 C_1. \tag{28}$$

We emphasize that constants C_1 and C_2 in (28) depend on μ_0, μ_1^0, μ_2^0 and R but do not depend on functions μ_1 and μ_2 satisfying condition (i). Let us formulate the results obtained as the following theorem.

Theorem 1 *Let conditions (i), (ii) take place. Then for any pair* $\mu = (\mu_1, \mu_2) \in L^\infty_{\mu_1^0}(\Omega_1) \times L^\infty_{\mu_2^0}(\Omega_2)$ *we have:*

(1) the operator $A_\mu : X \to X^*$*, defined in (26), is an isomorphism and for the inverse operator* $A_\mu^{-1} : X^* \to X$ *the estimate* $\|A_\mu^{-1}\| \le C_1 \equiv (1/\mu_*)$ *holds where constant* μ_* *is defined in (25);*
(2) for any external field $\Phi^e \in H(\Omega_3)$ *problem (15) has a unique solution* $\Phi_\mu \in X$*, and the estimate (28) holds where the constant* C_2 *depends on* μ_0, μ_1^0, μ_2^0 *and* R *but is independent of* (μ_1, μ_2).

Below, along with Eq. (27), we will consider an operator equation

$$A_\mu^* \Psi = \hat{F}, \quad \hat{F} \in X^*. \tag{29}$$

Here $A_\mu^* : X \to X^*$ is an adjoint operator for $A_\mu : X \to X^*$ which is defined by

$$\langle A_\mu^* \Psi, \Phi \rangle = a_\mu(\Phi, \Psi) = \langle A_\mu \Phi, \Psi \rangle \forall (\Phi, \Psi) \in X \times X.$$

From properties of adjoint operators and Theorem 1 follows:

Corollary 1 *Let the conditions of Theorem 1 hold. Then:*

(1) Operator $A_\mu^* : X \to X^*$ *which is adjoint for* A_μ *is isomorphism;*
(2) Equation (29) has a unique solution $\Psi \in X$ *for any element* $\hat{F} \in X^*$*, and the estimate* $\|\Psi\|_X \leq C_1 \|\hat{F}\|_{X^*}$ *holds for the solution.*

3 Formulation of the Inverse Problem. Applying the Optimization Method. Derivation of the Optimality System

The main goal of this paper is analysis of the inverse problem arising while developing the design technologies of magnetic bilayer cloaking devices. This problem consists of finding unknown permeabilities μ_1 and μ_2 of media filling Ω_1 and Ω_2, respectively, using given information about magnetic potential Φ in some subset $Q \subset B_R$. To achieve this goal we apply the optimization method. This approach is based on minimization of a cost functional which corresponds to the initial problem of designing approximate cloaking device. As a result the initial cloaking problem is reduced to studying corresponding control problem using the well-known methods of solving extremum problems. As a cost functional we choose one of the following:

$$I(\Phi) = \|\Phi - \Phi^d\|_Q^2 = \int_Q (\Phi - \Phi^d)^2 dx, \; I_2(\Phi) = \|\nabla\Phi + \mathbf{H}^d\|_Q^2. \qquad (30)$$

Here $Q \subset \Omega_0 \cup \Omega_3$ is a bounded open subset, Φ^d (or $\mathbf{H}^d$) is a given in Q function. One can read about another possible cost functionals in [19]. As controls we choose variable permeabilities μ_1 and μ_2. We will assume that the controls μ_1 and μ_2 are changed over certain sets K_1 and K_2 satisfying

(j) $K_1 \subset H^s_{\mu_1^0}(\Omega_1)$, $\mu_1^0 = \text{const} > 0$, $K_2 \subset H^s_{\mu_2^0}(\Omega_2)$, $\mu_2^0 = \text{const} > 0$, $s > 1$; $\alpha_0 > 0$.

Setting $K = K_1 \times K_2$, $\mu = (\mu_1, \mu_2)$ we define the operator $G : X \times K \to X^*$ by

$$\langle G(\Phi, \mu), S \rangle = \langle A_\mu \Phi, S \rangle - \langle F, S \rangle \equiv a_\mu(\Phi, S) - \langle F, S \rangle \;\; \forall S \in X \qquad (31)$$

and consider the following control problem:

$$J(\Phi, \mu) \equiv \frac{\alpha_0}{2} I(\Phi) + \frac{\alpha_1}{2} \|\mu_1\|_{s,\Omega_1}^2 + \frac{\alpha_2}{2} \|\mu_2\|_{s,\Omega_2}^2 \to \inf, \; G(\Phi, \mu) = 0, \; (\Phi, \mu) \in X \times K. \qquad (32)$$

Here I is a weakly lower semicontinuous functional, $\alpha_0, \alpha_1, \alpha_2$ are non-negative parameters which are used to control the relative importance of terms in (32). Let $Z_{ad} = \{(\Phi, \mu) \in X \times K : G(\Phi, \mu) = 0, J(\Phi, \mu) < \infty\}$.

We apply the mathematical procedure developed in [19] for studying control problems having the form (32). Based on this procedure, we can prove the following theorem.

Theorem 2 *Let, under assumptions (ii) and (j),* $\alpha_1 > 0$, $\alpha_2 > 0$ *or* $\alpha_1 \geq 0$, $\alpha_2 \geq 0$ *and* K_1, K_2 *be bounded sets. Let* $I : X \to R$ *be a weakly lower semicontinuous functional. Then control problem (32) has at least one solution* $(\Phi, \mu_1, \mu_2) \in X \times K_1 \times K_2$.

Proof Denote by $(\Phi^m, \mu^m) \in Z_{ad}$, $\mu^m = (\mu_1^m, \mu_2^m)$, $m \in \hat{N} \equiv \{1, 2, \dots\}$ a minimizing sequence for the functional J for which the following relations hold:

$$a_0(\Phi^m, S) + a(\mu_1^m, \mu_2^m; \Phi^m, S) = \langle F, S \rangle \ \forall S \in X, \quad (33)$$

$$\lim_{m\to\infty} J(\Phi^m, \mu_m) = \inf_{(\Phi,\mu)\in Z_{ad}} J(\Phi, \mu) \equiv J^*.$$

By conditions of Theorem 2 we have

$$\|\mu_1^m\|_{s,\Omega_1} \leq c_1, \quad \|\mu_2^m\|_{s,\Omega_2} \leq c_2 \ \forall m \in \hat{N}. \quad (34)$$

Here and below c_1, c_2, c_3 are some constants which do not depend on m. It follows from (34) and Theorem 1 that $\|\Phi^m\|_X \leq c_3$. From this estimate and (34) we derive that there are weak limits $\mu_1^* \in K_1$, $\mu_2^* \in K_2$, $\Phi^* \in X$ of some subsequences of sequences $\{\mu_1^m\}$, $\{\mu_2^m\}$, $\{\Phi^m\}$. Using this fact and compactness of embedding $H^s(\Omega_k) \subset L^\infty(\Omega_k)$ at $s > 1$, $k = 1, 2$, we conclude that $\mu_1^m \to \mu_1^*$ strongly in $L^\infty(\Omega_1)$, $\mu_2^m \to \mu_2^*$ strongly in $L^\infty(\Omega_2)$, while $\Phi^m \to \Phi^*$ weakly in X.

Let us prove that $G(\Phi^*, \mu_1^*, \mu_2^*) = 0$, i.e. that

$$a_0(\Phi^*, S) + a(\mu_1^*, \mu_2^*; \Phi^*, S) = \langle F, S \rangle \ \ \forall S \in X. \quad (35)$$

To this end we pass to the limit in (33) as $m \to \infty$. It is clear that linear term $a_0(\Phi^m, S)$ passes to the term $a_0(\Phi^*, S)$ as $m \to \infty$ while for the difference $a(\mu_1^m, \mu_2^m; \Phi^m, S) - a(\mu_1^*, \mu_2^*; \Phi^*, S)$ we have by (17)

$$|a(\mu_1^m, \mu_2^m; \Phi^m, S) - a(\mu_1^*, \mu_2^*; \Phi^*, S)| \leq$$

$$\leq |a_1(\mu_1^m; \Phi^m, S) - a_1(\mu_1^*; \Phi^*, S)| + |a_2(\mu_2^m; \Phi^m, S) - a_2(\mu_2^*; \Phi^*, S)|. \quad (36)$$

Let us show that every of terms in the right-hand side of (36) vanishes as $m \to \infty$. In fact, taking into consideration (18), we have for the first term in (36):

$$|a_1(\mu_1^m; \Phi^m, S) - a_1(\mu_1^*; \Phi^*, S)| \leq$$

$$\leq \left| \int_{\Omega_1} \mu_1^*(\nabla\Phi^m - \nabla\Phi^*) \cdot \nabla S d\mathbf{x} \right| + \left| \int_{\Omega_1} (\mu_1^m - \mu_1^*) \nabla\Phi^m \cdot \nabla S d\mathbf{x} \right| \ \forall S \in X. \quad (37)$$

As $\Phi^m \to \Phi^*$ weakly in X then the first integral in the right-hand side of (37) tends to zero as $m \to \infty$ for any $S \in X$. Besides, it follows from the strong convergence $\mu_1^m \to \mu_1^*$ in $L^\infty(\Omega_1)$ that the second integral in (37) also tends to zero as $m \to \infty$. This means that the first term in the right-hand side of (36) tends to zero as $m \to \infty$. In a similar way one can show that the second term in the right-hand side of (36) tends to zero as $m \to \infty$. Therefore passing to the limit in (33) as $m \to \infty$ we arrive at (35). This means that $G(\Phi^*, \mu^*) = 0$ where $\mu^* = (\mu_1^*, \mu_2^*)$. As the functional $J(\Phi)$ is a weakly lower semicontinuous functional on $X \times K$ we derive that $J(\Phi^*, \mu^*) = J^*$. This proves the theorem. □

Let us derive the necessary optimality conditions for problem (32). For this purpose we make use of the extremum principle in smoothly-convex extremum problems [32]. Preliminarily we find the Fréchet derivative with respect to Φ of the operator $G : X \times K \to X^*$ defined in (31). It follows from linearity of the operator G with respect to Φ that the Fréchet derivative $G'_\Phi(\hat{\Phi}, \hat{\mu})$ at every point $(\hat{\Phi}, \hat{\mu}) \in X \times K$ where $\hat{\mu} = (\hat{\mu}_1, \hat{\mu}_2)$ is defined by $G'_\Phi(\hat{\Phi}, \hat{\mu}) = \hat{A} \equiv A_{\hat{\mu}}$. Here operator $A_{\hat{\mu}}$ is defined by (26) at $\mu = \hat{\mu}$. Besides, it follows from (30) that

$$\langle (I_1')_\Phi(\hat{\Phi}), S\rangle = 2(\hat{\Phi} - \Phi^d, S)_Q =$$

$$= 2\int_Q (\hat{\Phi} - \Phi^d) S d\mathbf{x}, \ \langle (I_2)'_\Phi(\hat{\Phi}), S\rangle = 2(\nabla\hat{\Phi} + \mathbf{H}^d, \nabla S) \ \forall S \in X. \quad (38)$$

Following [32] we introduce a Lagrange multiplier $\Psi \in X$ which will be referred to as an adjoint magnetic field and consider the Lagrangian $\mathcal{L} : X \times K \times X \to R$ defined by the formula $L(\Phi, \mu, \Psi) \equiv J(\Phi, \mu) + \langle G(\Phi, \mu), \Psi\rangle_{X^* \times X}$. Denote by $\hat{A}^* \equiv G'_\Phi(\hat{\Phi}, \hat{\mu})^* : X \to X^*$ the operator adjoint for operator $\hat{A} = G'_\Phi(\hat{\Phi}, \hat{\mu}) : X \to X^*$ defined by

$$\langle \hat{A}^*\Psi, S\rangle_{X^* \times X} \equiv \langle G'_\Phi(\hat{\Phi}, \hat{\mu})^*\Psi, S\rangle_{X^* \times X} = \langle G'_\Phi(\hat{\Phi}, \hat{\mu})S, \Psi\rangle_{X^* \times X} =$$

$$\langle \hat{A}S, \Psi\rangle_{X^* \times X} \ \forall \Psi \in X, \ S \in X. \quad (39)$$

It follows from linearity of operator G with respect to μ_1, μ_2 and from convexity of the set $K = K_1 \times K_2$ that the set $G(\Phi, K) = \{\mathbf{x}^* = G(\Phi, \mu) \in X^*, \mu \in K\}$ is a convex subset of X^* for any function $\Phi \in X$. As the operator $G'_\Phi(\hat{\Phi}, \hat{\mu}) \equiv \hat{A}$ is an isomorphism by Theorem 1, then from results of [32, p. 83] the next theorem follows.

Theorem 3 *Let under assumptions (ii) and (j) the pair $(\hat{\Phi}, \hat{\mu}) \in X \times K$ where $\hat{\mu} = (\hat{\mu}_1, \hat{\mu}_2)$ be a solution of problem (32) and let functional $I(\Phi)$ be continuously differentiable with respect to state Φ in point $\hat{\Phi}$. Then there exists a unique Lagrange multiplier $\Psi \in X$ that satisfies the Euler-Lagrange equation*

$$a_{\hat{\mu}}(S, \Psi) \equiv a_0(S, \hat{\Psi}) + a_1(\hat{\mu}_1; S, \hat{\Psi}) + a_2(\hat{\mu}_2; S, \hat{\Psi}) = -(\alpha_0/2)\langle I'_\Phi(\hat{\Phi}), S\rangle \forall S \in X \quad (40)$$

and the minimum principle holds, which is equivalent to the following variational inequalities:

$$\alpha_1(\hat{\mu}_1, \mu_1 - \hat{\mu}_1)_{s,\Omega_1} + a_1\left(\mu_1 - \hat{\mu}_1, \hat{\Phi}, \hat{\Psi}\right) \geq 0 \; \forall \mu_1 \in K_1, \tag{41}$$

$$\alpha_2(\hat{\mu}_2, \mu_2 - \hat{\mu}_2)_{s,\Omega_2} + a_2\left(\mu_2 - \hat{\mu}_2, \hat{\Phi}, \hat{\Psi}\right) \geq 0 \; \forall \mu_2 \in K_2. \tag{42}$$

Direct problem (15), identity (40) which has the meaning of an adjoint problem for the adjoint state $\hat{\Psi} \in X$ and variational inequalities (41) and (42) constitute the optimality system. It describes the necessary conditions of an extremum for control problem (32). In particular case when $I = I_1(\Phi)$ we have by (38) that the Euler-Lagrange equation (40) has the form

$$a_0(S, \hat{\Psi}) + a_1(\hat{\mu}_1; S, \hat{\Psi}) + a_2(\hat{\mu}_2; S, \hat{\Psi}) = -\alpha_0(\hat{\Phi} - \Phi^d, S)_Q \forall S \in X. \tag{43}$$

Based on analysis of the optimality system we establish below sufficient conditions for the data which provide uniqueness and stability of solutions of problem (32) with respect to small disturbances of function $\Phi^d \in L^2(Q)$ entering into the cost functional in (30).

4 Uniqueness and Stability of Optimal Solutions

We will assume that the external field Φ^e can change in a bounded domain $\mathcal{H}_{ad} \in H(\Omega_3)$ and consider in this Section the control problem

$$J(\Phi, \mu) \equiv \frac{\alpha_0}{2}\|\Phi - \Phi^d\|_Q^2 + \frac{\alpha_1}{2}\|\mu_1\|_{s,\Omega_1}^2 + \frac{\alpha_2}{2}\|\mu_2\|_{s,\Omega_2}^2 \to \inf,$$
$$G(\Phi, \mu) = 0, \quad (\Phi, \mu) \in X \times K \tag{44}$$

corresponding to cost functional $I_1(\Phi) = \|\Phi - \Phi^d\|_Q^2$. Let us denote by $(\Phi_1, \mu_1) = (\Phi_1, \mu_1^{(1)}, \mu_2^{(1)})$ a solution of problem (44) corresponding to given function $\Phi_1^d = \Phi^d \in L^2(Q)$. By $(\Phi_2, \mu_2) = (\Phi_2, \mu_1^{(2)}, \mu_2^{(2)})$ we denote a solution of problem

$$J(\Phi, \mu) \equiv \frac{\alpha_0}{2}\|\Phi - \tilde{\Phi}^d\|_Q^2 + \frac{\alpha_1}{2}\|\mu_1\|_{s,\Omega_1}^2 + \frac{\alpha_2}{2}\|\mu_2\|_{s,\Omega_2}^2 \to \inf,$$
$$G(\Phi, \mu) = 0, \quad (\Phi, \mu) \in X \times K. \tag{45}$$

It is obtained from (44) by replacing functional $I(\Phi) = \|\Phi - \Phi^d\|_Q^2$ by functional $\tilde{I}(\Phi) = \|\Phi - \tilde{\Phi}^d\|_Q^2$, where $\tilde{\Phi}^d \in L^2(Q)$ is a function close to Φ^d in the norm of $L^2(Q)$. We note that by Theorem 1 the following estimate holds for Φ_l, $l = 1, 2$:

$$\|\Phi_l\|_X \leq M_\Phi \equiv C_2\|\Phi^e\|_{1,\Omega_3}. \tag{46}$$

Denote by $\Psi_l \in X,\ l = 1, 2$ Lagrange multipliers corresponding to solutions $(\Phi_l, \mu^{(l)})$. By Theorem 3 they satisfy

$$a_{\mu^{(l)}}(S, \Psi_l) \equiv a_0(S, \Psi_l) + a(\mu_1^{(l)}, \mu_2^{(l)}; S, \Psi_l) = -\alpha_0(\Phi_l - \Phi^d, S)_Q \ \forall S \in X,\ l = 1, 2. \quad (47)$$

We set $\mu = \mu_1 - \mu_2$,

$$\mu_1 = \mu_1^{(1)} - \mu_1^{(2)},\ \mu_2 = \mu_2^{(1)} - \mu_2^{(2)},\ \Phi = \Phi_1 - \Phi_2,\ \Psi = \Psi_1 - \Psi_2,\ \Phi^d = \Phi_1^d - \Phi_2^d. \quad (48)$$

The following Lemma holds.

Lemma 1 *Let in addition to conditions (ii), (j) the triples* $(\Phi_1, \mu_1^{(1)}, \mu_2^{(1)})$ *and* $(\Phi_2, \mu_1^{(2)}, \mu_2^{(2)})$ *be solutions of problems (44) and (45), respectively, where* $\Phi^d = \Phi_1^d, \tilde{\Phi}^d = \Phi_2^d \in L^2(Q)$ *are given functions. Let* $\Psi_l \in X,\ l = 1, 2$ *be the Lagrange multipliers corresponding to solutions* $(\Phi_l, \mu_1^{(l)}, \mu_2^{(l)})$. *Then the following estimate and inequality hold:*

$$\|\Phi\|_X \le C_1 C_s(\|\mu_1\|_{s,\Omega_1} + \|\mu_2\|_{s,\Omega_2}) M_\Phi, \quad (49)$$

$$\alpha_0(\Phi - \Phi^d, \Phi)_Q \le -a\,(\mu_1, \mu_2; \Phi, \Psi_1 + \Psi_2) - \alpha_1\|\mu_1\|_{s,\Omega_1}^2 - \alpha_2\|\mu_2\|_{s,\Omega_2}^2. \quad (50)$$

Proof We subtract identity (15) written for $(\Phi_2, \mu_1^{(2)}, \mu_2^{(2)})$ from (15) written for $(\Phi_1, \mu_1^{(1)}, \mu_2^{(1)})$. Using notation (48) we obtain the following identity for the difference $\Phi = \Phi_1 - \Phi_2$:

$$a_0(\Phi, S) + a(\mu_1^{(2)}, \mu_2^{(2)}; \Phi, S) = -a(\mu_1, \mu_2; \Phi_1, S)\ \ \forall S \in X. \quad (51)$$

Using estimates (9), (19) and (46) we have

$$|a(\mu_1, \mu_2; \Phi_1, S)| \le C_s(\|\mu_1\|_{s,\Omega_1} + \|\mu_2\|_{s,\Omega_2})\|\Phi_1\|_X\|S\|_X \le$$

$$\le C_s(\|\mu_1\|_{s,\Omega_1} + \|\mu_2\|_{s,\Omega_2}) M_\Phi \|S\|_X\ \ \forall S \in X. \quad (52)$$

Taking into account estimates (52) it follows from Theorem 1 applied to the linear problem with respect to the difference $\Phi = \Phi_1 - \Phi_2$ that the estimate (49) holds for the solution Φ.

Now, we prove (50). First of all we set $\mu_1 = \mu_1^{(1)}$ in inequality (41) written for $\hat{\mu}_1 = \mu_1^{(2)}, \hat{\Phi} = \Phi_2, \hat{\Psi} = \Psi_2$ and then we set $\mu_1 = \mu_1^{(2)}$ in (41) written for $\hat{\mu}_1 = \mu_1^{(1)}$, $\hat{\Phi} = \Phi_1, \hat{\Psi} = \Psi_1$. Using (48) we obtain the inequalities

$$\alpha_1(\mu_1^{(2)}, \mu_1)_{s,\Omega_1} + a_1\,(\mu_1; \Phi_2, \Psi_2) \ge 0,\ -\alpha_1(\mu_1^{(1)}, \mu_1)_{s,\Omega_1} - a_1\,(\mu_1; \Phi_1, \Psi_1) \ge 0. \quad (53)$$

Summing these inequalities we arrive at the following relation for the difference $\mu_1 \equiv \mu_1^{(1)} - \mu_1^{(2)}$:

$$a_1(\mu_1; \Phi_1, \Psi_1) - a_1(\mu_1; \Phi_2, \Psi_2) \le -\alpha_1 \|\mu_1\|^2_{s,\Omega_1}. \quad (54)$$

In the same manner, we obtain the second inequality for the difference $\mu_2 \equiv \mu_2^{(1)} - \mu_2^{(2)}$:

$$a_2(\mu_2; \Phi_1, \Psi_1) - a_2(\mu_2; \Phi_2, \Psi_2) \le -\alpha_2 \|\mu_2\|^2_{s,\Omega_2}. \quad (55)$$

We subtract identity (47) for $l = 2$ from (47) for $l = 1$ and set $S = \Phi$. We obtain

$$a_0(\Phi, \Psi) + a(\mu_1^{(2)}, \mu_2^{(2)}; \Phi, \Psi) + a(\mu_1, \mu_2; \Phi, \Psi_1) = -\alpha_0(\Phi - \Phi^d, \Phi)_Q. \quad (56)$$

Then we subtract identity (51) for $S = \Phi$ from (56) and obtain

$$a(\mu_1, \mu_2; \Phi, \Psi_1) - a(\mu_1, \mu_2; \Phi_1, \Psi) = -\alpha_0(\Phi - \Phi^d, \Phi)_Q. \quad (57)$$

Summing (57) with (54) and (55) we arrive taking unto account (17) at the desired inequality (50) after simple transformations. The lemma is proved. □

Setting $M_\Phi^0 = C_P M_\Phi + \max(\|\Phi_1^d\|_Q, \|\Phi_2^d\|_Q)$ where C_P is a constant entering into Poincaré-Friedrichs inequality (13) we are able to prove the following main theorem.

Theorem 4 *Let, in addition to assumptions (ii), (j),* $K = K_1 \times K_2 \subset H^s_{\mu_1^0}(\Omega_1) \times H^s_{\mu_2^0}(\Omega_2)$ *be a bounded set and let a triple* $(\Phi_l, \mu_1^{(l)}, \mu_2^{(l)})$ *be a solution of control problem (44) corresponding to a given function* $\Phi_l^d = \Phi^d \in L^2(Q)$, $l = 1, 2$, *where* $Q \subset \Omega_0 \cup \Omega_3$ *is a nonempty open subset. We assume that the following conditions are satisfied:*

$$\alpha_1(1-\varepsilon) > 4\alpha_0 C_1^2 C_s^2 M_\Phi M_\Phi^0, \quad \alpha_2(1-\varepsilon) > 4\alpha_0 C_1^2 C_s^2 M_\Phi M_\Phi^0 \quad (58)$$

where $\varepsilon \in (0, 1)$ *is an arbitrary constant. Then the following stability estimates hold:*

$$\|\mu_1^{(1)} - \mu_1^{(2)}\|_{s,\Omega_1} \le (1/2)\sqrt{\alpha_0/\varepsilon\alpha_1}\|\Phi_1^d - \Phi_2^d\|_Q, \quad (59)$$

$$\|\mu_2^{(1)} - \mu_2^{(2)}\|_{s,\Omega_2} \le (1/2)\sqrt{\alpha_0/\varepsilon\alpha_2}\|\Phi_1^d - \Phi_2^d\|_Q, \quad (60)$$

$$\|\Phi_1 - \Phi_2\|_Q \le \|\Phi_1^d - \Phi_2^d\|_Q, \quad (61)$$

$$\|\Phi_1 - \Phi_2\|_X \le (1/2)C_1 C_s(\sqrt{\alpha_0/\varepsilon\alpha_1} + \sqrt{\alpha_0/\varepsilon\alpha_2})\|\Phi_1^d - \Phi_2^d\|_Q. \quad (62)$$

Proof First, we derive the estimate of the Lagrange multiplier Ψ_l corresponding to the triple $(\Phi_l, \mu_1^{(l)}, \mu_2^{(l)})$, $l = 1, 2$. For this purpose, we turn to the problem (47), which is equivalent to the following operator equation for the multiplier Ψ_l:

$$A^*_{\mu^{(l)}}\Psi_l = \alpha_0 f_l \in X^*, \quad \langle f_l, S\rangle = -(\Phi_l - \Phi_l^d, S)_Q, \quad l = 1, 2. \tag{63}$$

Here $A^*_{\mu^{(l)}}$ is the adjoint for the operator $A_{\mu^{(l)}}$, defined by (26) for $\mu = \mu^{(l)} = (\mu_1^{(l)}, \mu_2^{(l)})$, $l = 1, 2$. It is easy to see that

$$|(\Phi_l - \Phi_l^d, S)_Q| \leq [\|\Phi_l\|_Q + \max(\|\Phi_1^d\|_Q, \|\Phi_2^d\|_Q)]\|S\|_X \leq$$

$$\leq [C_P\|\Phi_l\|_X + \max(\|\Phi_1^d\|_Q, \|\Phi_2^d\|_Q)]\|S\|_X = M_\Phi^0\|S\|_X \;\forall S \in X.$$

From this fact it follows that $\|f_l\|_{X^*} \leq M_\Phi^0$, and it follows from Corollary 1 applied to problem (63) for Ψ_l that the following estimate holds for Ψ_l:

$$\|\Psi_l\|_X \leq C_1\alpha_0\|f_l\|_{X^*} \leq C_1\alpha_0 M_\Phi^0, \quad M_\Phi^0 \equiv C_P M_\Phi + \max(\|\Phi_1^d\|_Q, \|\Phi_2^d\|_Q). \tag{64}$$

Combining estimates (9), (19), (49), (64) and the Young inequality $2cd \leq \varepsilon c^2 + (1/\varepsilon)d^2\;\forall c \geq 0, d \geq 0, \varepsilon > 0$ at $\varepsilon = 1$, we derive using conditions (58) that

$$\begin{aligned}
&|a(\mu_1, \mu_2; \Phi, \Psi_1 + \Psi_2)| \leq C_s(\|\mu_1\|_{s,\Omega_1} + \|\mu_2\|_{s,\Omega_2})\|\Phi\|_X(\|\Psi_1\|_X + \|\Psi_2\|_X) \leq \\
&\leq 2C_1\alpha_0 M_\Phi^0 C_s(\|\mu_1\|_{s,\Omega_1} + \|\mu_2\|_{s,\Omega_2})\left[C_1 C_s(\|\mu_1\|_{s,\Omega_1} + \|\mu_2\|_{s,\Omega_2})M_\Phi\right] = \\
&\leq 4\alpha_0 C_1^2 C_s^2 M_\Phi M_\Phi^0(\|\mu_1\|^2_{s,\Omega_1} + \|\mu_2\|^2_{s,\Omega_2}) \leq \\
&\leq \alpha_1(1-\varepsilon)\|\mu_1\|^2_{s,\Omega_1} + \alpha_2(1-\varepsilon)\|\mu_2\|^2_{s,\Omega_2}.
\end{aligned} \tag{65}$$

Using (65) we deduce from (50) that

$$\alpha_0\|\Phi\|_Q^2 \leq \alpha_0\|\Phi^d\|_Q\|\Phi\|_Q - \varepsilon\alpha_1\|\mu_1\|^2_{s,\Omega_1} - \varepsilon\alpha_2\|\mu_2\|^2_{s,\Omega_2}. \tag{66}$$

From (66) we obtain the estimate $\|\Phi\|_Q \leq \|\Phi^d\|_Q$, which is equivalent by (48) to estimate (61). Next, using given estimate and the inequality $\|\Phi^d\|_Q\|\Phi\|_Q \leq \|\Phi\|_Q^2 + (1/4)\|\Phi^d\|_Q^2$, which follows from the Young inequality at $\varepsilon = 1/2$, we derive from (66) that

$$\varepsilon\alpha_1\|\mu_1\|^2_{s,\Omega_1} + \varepsilon\alpha_2\|\mu_2\|^2_{s,\Omega_2} \leq \alpha_0\|\Phi^d\|_Q\|\Phi\|_Q - \alpha_0\|\Phi\|_Q^2 \leq \frac{\alpha_0}{4}\|\Phi^d\|_Q^2. \tag{67}$$

From (67) we obtain estimates (59), (60) for $\|\mu_1^{(1)} - \mu_1^{(2)}\|_{s,\Omega_1}$ and $\|\mu_2^{(1)} - \mu_2^{(2)}\|_{s,\Omega_2}$, while the estimate (62) follows from estimates (49), (59) and (60). The theorem is proved. □

Note that the stability of the solution of extremum problem (32) has been proved assuming that the parameters α_1 and α_2 are positive. This means that the terms

$(\alpha_1/2)\|\mu_1\|^2_{s,\Omega_1}$ and $(\alpha_2/2)\|\mu_2\|^2_{s,\Omega_2}$ in the expression for the functional J in (32) have a regularizing effect.

In a similar way one can prove the stability of solutions of control problem

$$J(\Phi,\mu)\equiv\frac{\alpha_0}{2}\|\nabla\Phi+\mathbf{H}^d\|^2_Q+\frac{\alpha_1}{2}\|\mu_1\|^2_{s,\Omega_1}+\frac{\alpha_2}{2}\|\mu_2\|^2_{s,\Omega_2}\to\inf,$$

$$G(\Phi,\mu)=0,\quad(\Phi,\mu)\in X\times K \tag{68}$$

corresponding to the second functional $I_2(\Phi)$ in (30).

Theorem 5 *Let, in addition to assumptions (ii), (j), $K=K_1\times K_2$ be a bounded set and let a triple $(\Phi_l,\mu_1^{(l)},\mu_2^{(l)})$ be a solution of control problem (68) corresponding to a given function $\mathbf{H}^d=\mathbf{H}_l^d\in L^2(Q)$, $l=1,2$, where $Q\subset\Omega_0\cup\Omega_3$ is a nonempty open subset. Let $M_\Phi^0=M_\Phi+\max(\|\mathbf{H}_1^d\|_Q,\|\mathbf{H}_2^d\|_Q)$ where M_Φ is defined in (46) and assume that conditions (58) take place. Then the following stability estimates hold:*

$$\|\mu_1^{(1)}-\mu_1^{(2)}\|_{s,\Omega_1}\le(1/2)\sqrt{\alpha_0/\varepsilon\alpha_1}\|\mathbf{H}_1^d-\mathbf{H}_2^d\|_Q, \tag{69}$$

$$\|\mu_2^{(1)}-\mu_2^{(2)}\|_{s,\Omega_2}\le(1/2)\sqrt{\alpha_0/\varepsilon\alpha_2}\|\mathbf{H}_1^d-\mathbf{H}_2^d\|_Q, \tag{70}$$

$$\|\Phi_1-\Phi_2\|_Q\le\|\mathbf{H}_1^d-\mathbf{H}_2^d\|_Q, \tag{71}$$

$$\|\Phi_1-\Phi_2\|_X\le(1/2)C_1C_s(\sqrt{\alpha_0/\varepsilon\alpha_1}+\sqrt{\alpha_0/\varepsilon\alpha_2})\|\mathbf{H}_1^d-\mathbf{H}_2^d\|_Q. \tag{72}$$

In conclusion, we studied control problems for magnetic scattering model (2)–(4). We studied some new properties of solutions of direct scattering problem, proved the correctness of this problem and solvability of control problem (32) and derived the optimality system (15), (41)–(43) describing the necessary conditions of extremum. Based on analysis of the optimality system, we have established sufficient conditions for the initial data which provide the uniqueness and stability of solutions of problems (44) and (68) with respect to small perturbations of functions $\Phi^d\in L^2(Q)$ or $\mathbf{H}^d\in L^2(Q)$ included in the respective cost functionals in (32). In the same manner, the stability of optimal solutions with respect to small perturbations of the external field Φ^e can be investigated.

It should also be noted that the optimality system, which has been derived above, can be used to develop an efficient numerical algorithm for solving the extremum problems (32) and (68). An alternative algorithm is based on the use of the particle swarm method [33] according to the scheme proposed in [34, 35] for the numerical solution of the thermal cloaking problems of a cylindrical and spherical bodies. Authors plan to devote a forthcoming paper to developing and studying the properties of the algorithms and to comparative analysis of results of numerical experiments performed using these algorithms.

Acknowledgements The first author was supported by the Russian Science Foundation (project no. 14-11-00079). The second author acknowledge the support by the Russian Foundation for Basic Research (project no. 16-01-00365-a).

References

1. L.S. Dolin, On a possibility of comparison of three-dimensional electromagnetic systems with nonuniform anisotropic filling. Izv Vuzov Radiofizika **4**, 964–967 (1961)
2. J.B. Pendry, D. Shurig, D.R. Smith, Controlling electromagnetic fields. Science **312**, 1780–1782 (2006)
3. U. Leonhardt, Optical conformal mapping. Science **312**, 1777–1780 (2006)
4. A. Alú, N. Engheta, Achieving transparency with plasmonic and metamaterial coatings. Phys. Rev. E **72**, 016623 (2005)
5. S.A. Cummer, D. Shurig, One path to acoustic cloaking. New J. Phys. **9**, 45–51 (2007)
6. H. Chen, C.T. Chan, Acoustic cloaking in three dimensions using acoustic metamaterials. Appl. Phys. Lett. **91**, 183518 (2007)
7. B. Wood, J.B. Pendry, Metamaterials at zero frequency. J. Phys.: Condens. Matter. **19**, 076208 (2007)
8. S. Guenneau, C. Amra, D. Veynante, Transformation thermodynamics: cloaking and concentrating heat flux. Opt. Express **20**, 8207 (2012)
9. F. Yang, Z.L. Zhong Mei, T.Y. Jin, T.J. Cui, DC electric invisibility cloak. Phys. Rev. Lett. **109**, 053902 (2012)
10. S.V. Yampolskii, Y.A. Genenko, Magnetic cloaking by a paramagnet/superconductor cylindrical tube in the critical state. Appl. Phys. Lett. **104**, 143504 (2014)
11. T. Han, C.-W. Qiu, Transformation Laplacian metamaterials: recent advances in manipulating thermal and dc fields. J. Opt. **18**, 044003 (2016)
12. M. Raza, Y. Liu, E.H. Lee, Y. Yungui Ma, Transformation thermodynamics and heat cloaking: a review. J. Opt. **18**, 044002 (2016)
13. L. Kroon, K. Jarrendahl, Neutral shielding and cloaking of magnetic fields using isotropic media. J. Phys.: Condens. Matter **29**, 035801 (2017)
14. A. Sanchez, C. Navau, J. Prat-Camps, D.-X. Chen, Antimagnets: controlling magnetic fields with superconductor metamaterial hybrids. New J. Phys. **13**, 093034 (2011)
15. F. Gomory, M. Solovyov, J. Souc et al., Experimental realization of a magnetic cloak. Science **335**, 1466–1468 (2012)
16. F. Gomory, M. Solovyov, J. Souc et al., Supporting online materials for experimental realization of a magnetic cloak. Science **335**, 1466–1468 (2012)
17. G.V. Alekseev, V.G. Romanov, One class of nonscattering acoustic shells for a model of anisotropic acoustics. J. Appl. Ind. Math. **6**, 1–5 (2012)
18. S. Xu, Y. Wang, B. Zhang, H. Chen, Invisibility cloaks from forward design to inverse design. Sci. China Inf. Sci. **56**, 120408 (2013)
19. G.V. Alekseev, *Invisibility problem in acoustics, optics and heat transfer* (Dalnauka, Vladivostok, 2016). [in Russian]
20. B.-I. Popa, S.A. Cummer, Cloaking with optimized homogeneous anisotropic layers. Phys. Rev. A **79**, 023806 (2009)
21. X.H. Wang, E.A. Semouchkina, A route for efficient non-resonance cloaking by using multilayer dielectric coating. Appl. Phys. Lett. **102**, 113506 (2013)
22. A. Mirzaei, A.E. Miroshnichenko, I.V. Shadrivov, Y.S. Kivshar, All-dielectric multilayer cylindrical structures for invisibility cloaking. Sci. Rep. **5**, 9574 (2015)
23. R.V. Brizitskii, A.S. Savenkova, Inverse extremum problems for Maxwell's equations. Comput. Math. Math. Phys. **50**, 984–992 (2010)

24. G.V. Alekseev, R.V. Brizitskii, The theoretical analysis of boundary control extremal problems for Maxwell's equations. J. Appl. Ind. Math. **5**, 1–15 (2011)
25. G.V. Alekseev, Cloaking of material objects by controlling the impedance boundary condition for Maxwell's equations. Dokl. Phys. **58**, 482–486 (2013)
26. G.V. Alekseev, Cloaking via impedance boundary condition for 2-D Helmholtz equation. Appl. Anal. **93**, 254–268 (2014)
27. G.V. Alekseev, V.A. Levin, Optimization method of searching parameters of an inhomogeneous liquid medium in the acoustic cloaking problem. Dokl. Phys. **59**, 89–93 (2014)
28. G.V. Alekseev, Stability estimates in the problem of cloaking material bodies for Maxwell's equations. Comp. Math. Mathem. Phys. **54**, 1788–1803 (2014)
29. G.V. Alekseev, Analysis and optimization in problems of cloaking of material bodies for the Maxwell equations. Differ. Eqn. **52**, 366–377 (2016)
30. D.S. Anikonov, V.G. Nazarov, I.V. Prokhorov, Visible and invisible media in tomography. Dokl. Math. **56**, 955–958 (1997)
31. D.S. Anikonov, V.G. Nazarov, I.V. Prokhorov, Poorly visible media in X-ray tomography, in *Inverse and Ill-Posed Problems Series*, vol. 38 (VSP, Utrecht, 2002), viii+296
32. A.V. Fursikov, *Optimal Control of Distributed Systems: Theory and Applications* (American Mathematical Society, Boston, USA, 2000)
33. R. Poli, J. Kennedy, T. Blackwell, Particle swarm optimization: an overview. Swarm Intel. **1**, 33–57 (2007)
34. G.V. Alekseev, A.V. Lobanov, Yu. E. Spivak, Optimization and discretization in 2D problems of electromagnetic invisible cloaking. CEUR Workshop Proc. **1623**, 125–137 (2016)
35. G.V. Alekseev, V.A. Levin, D.A. Tereshko, Optimization analysis of the thermal cloaking problem for a cylindrical body. Dokl. Phys. **62**, 71–75 (2017)

Synthesis Problem and Mathematical Modeling of Multilayered Absorbing Coating

E. D. Derevyanchuk, A. S. Shutkov and Yu. G. Smirnov

Abstract The paper is devoted to application of the waveguide method in mathematical modeling of multilayered radar absorbing coating. Mathematically, it leads to a synthesis problem. We present a numerical-analytical method for solution of this problem and show computational results which can be applied in design of multilayered coatings.

Keywords Synthesis problem · Reflection coefficient · Complex permittivity
Thickness of layers

1 Introduction

The synthesis problem of coatings is known for a long time. However, in recent decades, new materials have appeared that make it possible to achieve qualitatively better results in coating design. Among new materials there are metamaterials, and films based on graphene, and multi-layered anisotropic materials [4–9].

The coating technology has also changed significantly: today it is possible to obtain ultra-thin multilayer materials. Often, a synthesis problem is reduced to solving a complicated nonlinear problem of electrodynamics. Therefore, it is necessary to develop effective mathematical methods, computational algorithms, program codes, including parallel algorithms that can be implemented on supercomputers. All these factors will allow one to solve the synthesis problems at the modern level, which is necessary for applications.

E. D. Derevyanchuk (✉) · A. S. Shutkov · Yu. G. Smirnov
Penza State University, 40, Krasnaya street, Penza 440026, Russia
e-mail: katyader11@yandex.ru

A. S. Shutkov
e-mail: einstein9@rambler.ru

Yu. G. Smirnov
e-mail: smirnovyug@mail.ru

L. Beilina and Yu. G. Smirnov (eds.), *Nonlinear and Inverse Problems in Electromagnetics*, Springer Proceedings in Mathematics & Statistics 243,
https://doi.org/10.1007/978-3-319-94060-1_2

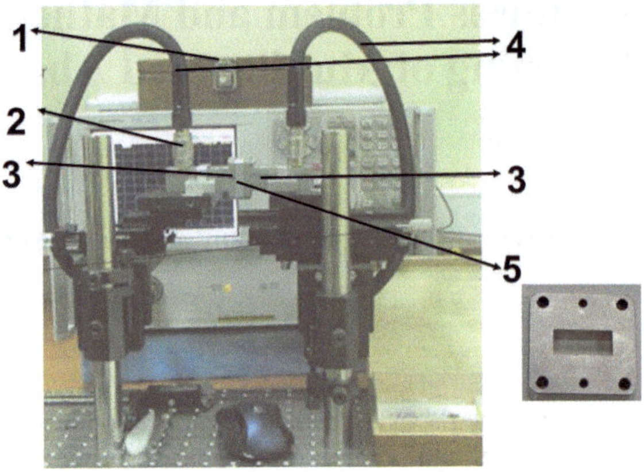

Fig. 1 Network analyser

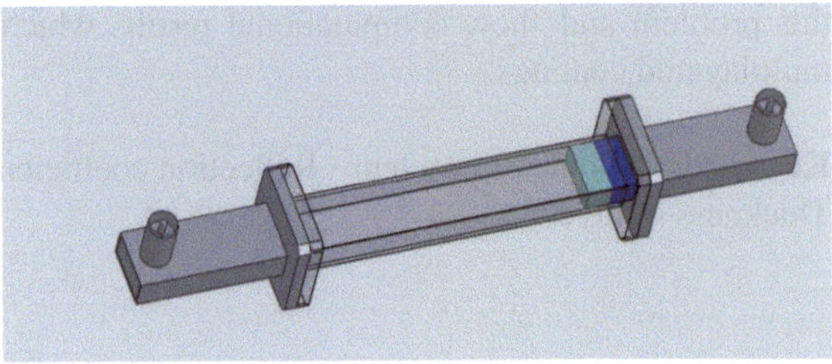

Fig. 2 The sample of the material is in the waveguide

The aim of the work is to develop the method and computational algorithm for solving such a problem. The synthesis problem is a complex nonlinear multiparameter problem that must be solved with high accuracy. To test multilayered coating the waveguide method is used [1]. For waveguide method we use network analyser (Fig. 1).

The material in the study (sample of the material) is put in the rectangular waveguide (Fig. 2). Electromagnetic field propagates in the waveguide. Measured data—amplitude and phase of reflection or transmission coefficient—are used both in inverse and synthesis problems.

2 Synthesis Problem

Assume that a waveguide $P = \{x : 0 < x_1 < a, 0 < x_2 < b, -\infty < x_3 < \infty\}$ with the perfectly conducting boundary surface ∂P is placed in the cartesian coordinates.

A multi-sectional diaphragm Q is placed in the waveguide (adjacent to the waveguide walls) (Fig. 3).

Section j of the diaphragm is described as follows

$$\Sigma_j = \{(x_1, x_2, x_3) : 0 \le x_1 \le a, 0 \le x_2 \le b, l_{j-1} \le x_3 \le l_j\}, j = 1, \ldots, n$$

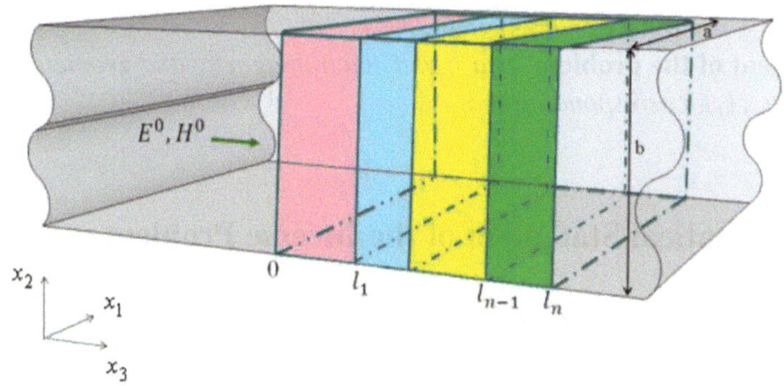

Fig. 3 Multi-sectional diaphragm in the waveguide

Note that $l_0 = 0$. Domain $P \backslash \bar{Q}$ is filled with an isotropic and homogeneous medium having constant permeability ($\mu_0 = 1$) and constant permittivity ($\varepsilon_0 = 1$).

The diaphragm has constant permeability μ_0. Each section Σ_j is filled with isotropic and homogeneous medium having constant permittivity:

$$\varepsilon_j(\omega) = \varepsilon_j^{'} - i\frac{\sigma_j}{\omega} \tag{1}$$

where ε_j^1 is a real part of the complex permittivity and σ_j is the conductivity ($j = 1, 2, \ldots, n$).

Introduce the notation $\underline{\boldsymbol{\varepsilon}} = \left(\underline{\varepsilon}_1, \ldots, \underline{\varepsilon}_n\right)^T$, $\varepsilon' = \left(\varepsilon_1', \ldots, \ldots, \varepsilon_n'\right)^T$, $\overline{\boldsymbol{\varepsilon}} = (\overline{\varepsilon}_1, \ldots, \overline{\varepsilon}_n)^T$, $\underline{\boldsymbol{\sigma}} = \left(\underline{\sigma}_1, \ldots, \underline{\sigma}_n\right)^T$, $\sigma' = \left(\sigma_1', \ldots, \ldots, \sigma_n'\right)^T$, $\overline{\boldsymbol{\sigma}} = (\overline{\sigma}_1, \ldots, \overline{\sigma}_n)^T$, where $4n$ constants $\underline{\varepsilon}_j$, $\overline{\varepsilon}_j$, $\underline{\sigma}_j$, $\overline{\sigma}_j$ are such that: $\underline{\varepsilon}_j \le \varepsilon_j' \le \overline{\varepsilon}_j'$ and $\underline{\sigma}_j \le \sigma_j \le \bar{\sigma}_j$, ($j = 1, 2, \ldots, n$).

Similarly, we introduce notations for the vector of thickness: $\underline{\boldsymbol{l}} \le \boldsymbol{l} \le \overline{\boldsymbol{l}}$, $\boldsymbol{l} = \left(l_1, l_2, \ldots, l_j, \ldots, l_n\right)^T$. We will use below vector inequalities $\underline{\varepsilon} \le \varepsilon' \le \overline{\varepsilon}$, $\underline{\sigma} \le \sigma \le \bar{\sigma}$, $\underline{\boldsymbol{l}} \le \boldsymbol{l} \le \overline{\boldsymbol{l}}$. We denote by $B(\omega, \boldsymbol{\varepsilon}', \boldsymbol{\sigma}, \boldsymbol{l})/A$ a value of reflection coefficient for the fixed vaslues of $\omega, \boldsymbol{\varepsilon}', \boldsymbol{\sigma}, \boldsymbol{l}$.

We will consider the following optimization problem in the range of circular frequencies $\omega_1 \le \omega \le \omega_2$:

$$\beta(\boldsymbol{\varepsilon}', \boldsymbol{\sigma}, \boldsymbol{l}) = \max_{\omega_1 \le \omega \le \omega_2} \left| B(\omega, \boldsymbol{\varepsilon}', \boldsymbol{\sigma}, \boldsymbol{l})/A \right|. \tag{2}$$

$$\eta = \min_{\underline{\varepsilon} \le \varepsilon' \le \bar{\varepsilon}, \underline{\sigma} \le \sigma \le \bar{\sigma}, \underline{l} \le l \le \bar{l}} \max_{\omega_1 \le \omega \le \omega_2} \left| B(\omega, \boldsymbol{\varepsilon}', \boldsymbol{\sigma}, \boldsymbol{l})/A \right| \tag{3}$$

or

$$\eta = \min_{\underline{\varepsilon} \le \varepsilon' \le \bar{\varepsilon}, \underline{\sigma} \le \sigma \le \bar{\sigma}, \underline{l} \le l \le \bar{l}} \beta(\boldsymbol{\varepsilon}', \boldsymbol{\sigma}, \boldsymbol{l}), \tag{4}$$

Here B/A is a reflection coefficient.

Statement of the problem: *find* η *and electromagnetic and geometrical parameters* ($\varepsilon'_j, \sigma_j, l_j$)*, corresponding to* η.

3 Mathematical Statement of the Inverse Problem

The electromagnetic field $\boldsymbol{E}$, $\boldsymbol{H}$ inside and outside the object in the waveguide is governed by Maxwell's equations

$$\begin{cases} \operatorname{rot} \mathbf{H} = -i\omega\varepsilon\mathbf{E} \\ \operatorname{rot} \mathbf{E} = i\omega\mu_0\mathbf{H}, \end{cases} \tag{5}$$

where ω is the circular frequency.

Assume that $\pi/a < k_0 < \pi/b$, where $k_0^2 = \omega^2\varepsilon_0\mu_0$ (a single-mode waveguide). The incident electrical field is

$$\mathbf{E}^0 = A \sin\left(\frac{\pi x_1}{a}\right) e^{-i\gamma_0 x_3} \tag{6}$$

with A is a known constant and $\gamma_0 = \sqrt{k_0^2 - \pi^2/a^2}$.

Solving the direct problem for Maxwell's equations, taking into account (1), we obtain explicit expressions for the field inside every section of the diaphragm and outside the diaphragm:

$$\begin{aligned} E^{(0)} &= \sin\left(\frac{\pi x_1}{a}\right)(Ae^{-i\gamma_0 x_3} + Be^{i\gamma_0 x_3}), \\ E^{(j)} &= \sin\left(\frac{\pi x_1}{a}\right)(C_j e^{-i\gamma_j x_3} + D_j e^{i\gamma_j x_3}), \\ E^{(n+1)} &= \sin\left(\frac{\pi x_1}{a}\right) F e^{-i\gamma_{n+1} x_3}, \quad \gamma_{n+1} = \gamma_0 \end{aligned} \tag{7}$$

$$j = 1, 2, \ldots, n.$$

Substituting (1) into Maxwell's equations we get:

$$\gamma_j = \sqrt{\omega^2\varepsilon_j\mu_0 - \frac{\pi^2}{a^2}}. \tag{8}$$

The conditions on the boundary surfaces $L := \{x_3 = 0, \ldots, x_3 = l_j, \ldots, x_3 = l_{n-1}\}$ of the diaphragm sections are

$$[E_y]|_L = 0, \qquad [H_x]|_L = 0. \tag{9}$$

Square brackets [·] denote the jump function via the boundary surfaces.

On the boundary surface $\{x_3 = l_n\}$ of conductor we have:

$$E_y|_{x_3=l_n} = 0 \; or \; E_y|_{x_3=l_n} = -\xi H, \tag{10}$$

where the first condition corresponds to a perfect conductor, the second one corresponds to a "good" conductor. Here $\xi = \sqrt{\frac{\mu_{n+1}}{\varepsilon_{n+1}}}$ is surface impedance of the medium in $n+1$ section. The synthesis problem can be reduced to solving the following nonlinear equation. From Maxwell's equations we have

$$|B/A| = \left| \frac{\gamma_{n-1} p_n^{(-)} + \gamma_n q_n^{(-)} \frac{Te^{i\gamma_n(2l_n - l_{n-1})} + e^{i\gamma_n l_{n-1}}}{Te^{i\gamma_n(2l_n - l_{n-1})} - e^{i\gamma_n l_{n-1}}}}{\gamma_{n-1} p_n^{(+)} - \gamma_n q_n^{(+)} \frac{Te^{i\gamma_n(2l_n - l_{n-1})} + e^{i\gamma_n l_{n-1}}}{Te^{i\gamma_n(2l_n - l_{n-1})} - e^{i\gamma_n l_{n-1}}}} \right|, \tag{11}$$

$$\begin{pmatrix} p_1 \\ q_1 \end{pmatrix} = \begin{pmatrix} 1 \\ 1 \end{pmatrix}, \begin{pmatrix} p_2^{(\pm)} \\ q_2^{(\pm)} \end{pmatrix} = \begin{pmatrix} \gamma_{j-1} p_1 \cos\alpha_j \pm \gamma_j q_1 i \sin\alpha_j \\ \gamma_{j-1} p_1 i \sin\alpha_j \pm \gamma_j q_1 \cos\alpha_j \end{pmatrix}, \tag{12}$$

$$\begin{pmatrix} p_{j+1}^{(\pm)} \\ q_{j+1}^{(\pm)} \end{pmatrix} = \begin{pmatrix} \gamma_{j-1} p_j^{(\pm)} \cos\alpha_j + \gamma_j q_j^{(\pm)} i \sin\alpha_j \\ \gamma_{j-1} p_j^{(\pm)} i \sin\alpha_j + \gamma_j q_j^{(\pm)} \cos\alpha_j \end{pmatrix}, \; j = \left(\overline{1, n}\right),$$

$$\alpha_j = \gamma_j (l_{j+1} - l_j), \quad \gamma_j = \sqrt{\omega^2 \varepsilon_j \mu_0 - \frac{\pi^2}{a^2}}$$

$$T = \frac{\frac{\xi}{i\omega}\gamma_n - 1}{\frac{\xi}{i\omega}\gamma_n + 1} \tag{13}$$

or

$$T = -1, \; \xi = 0. \tag{14}$$

Note that formulas (11)–(13) corresponds to a "good" conductor and (11)–(12), (14) corresponds to a "perfect" conductor.

3.1 A Perfect Conductor. Two-Sectional Diaphragm

Let us consider the case with a perfect conductor.

Based on Eqs. (11)–(12), (14) for a perfect conductor we have

$$\frac{B}{A} = \frac{\left\{\frac{\gamma_0}{\mu_0}\gamma_1\left(1+e^{-2i\gamma_1 l_1}\right)-\gamma_1^2\left(1-e^{-2i\gamma_1 l_1}\right)\right\}\left(e^{-i\gamma_2(2l_2-l_1)}-e^{-i\gamma_2 l_1}\right)}{\left\{\frac{\gamma_0}{\mu_0}\gamma_1\left(1+e^{-2i\gamma_1 l_1}\right)+\gamma_1^2\left(1-e^{-2i\gamma_1 l_1}\right)\right\}\left(e^{-i\gamma_2(2l_2-l_1)}-e^{-i\gamma_2 l_1}\right)} - \frac{\left\{\frac{\gamma_0}{\mu_0}\gamma_2\left(1-e^{-2i\gamma_1 l_1}\right)-\gamma_1\gamma_2\left(1+e^{-2i\gamma_1 l_1}\right)\right\}\left(e^{-i\gamma_2(2l_2-l_1)}+e^{-i\gamma_2 l_1}\right)}{\left\{\frac{\gamma_0}{\mu_0}\gamma_2\left(1-e^{-2i\gamma_1 l_1}\right)+\gamma_1\gamma_2\left(1+e^{-2i\gamma_1 l_1}\right)\right\}\left(e^{-i\gamma_2(2l_2-l_1)}+e^{-i\gamma_2 l_1}\right)} \qquad (15)$$

3.2 A "Good" Conductor. Two-Sectional Diaphragm

In the case of a "good" conductor we have:

$$\frac{B}{A} = \frac{\left\{\frac{\gamma_0}{\mu_0}\gamma_1\left(1+e^{-2i\gamma_1 l_1}\right)-\gamma_1^2\left(1-e^{-2i\gamma_1 l_1}\right)\right\}\left(e^{-i\gamma_2(2l_2-l_1)}+Se^{-i\gamma_2 l_1}\right)}{\left\{\frac{\gamma_0}{\mu_0}\gamma_1\left(1+e^{-2i\gamma_1 l_1}\right)+\gamma_1^2\left(1-e^{-2i\gamma_1 l_1}\right)\right\}\left(e^{-i\gamma_2(2l_2-l_1)}+Se^{-i\gamma_2 l_1}\right)} - \frac{\left\{\frac{\gamma_0}{\mu_0}\gamma_2\left(1-e^{-2i\gamma_1 l_1}\right)-\gamma_1\gamma_2\left(1+e^{-2i\gamma_1 l_1}\right)\right\}\left(e^{-i\gamma_2(2l_2-l_1)}-Se^{-i\gamma_2 l_1}\right)}{\left\{\frac{\gamma_0}{\mu_0}\gamma_2\left(1-e^{-2i\gamma_1 l_1}\right)+\gamma_1\gamma_2\left(1+e^{-2i\gamma_1 l_1}\right)\right\}\left(e^{-i\gamma_2(2l_2-l_1)}-Se^{-i\gamma_2 l_1}\right)} \qquad (16)$$

where for a "good" conductor (compare with (10)):

$$S = \frac{i\xi}{\omega\mu_0}, \quad \xi = \sqrt{\frac{\mu_{n+1}}{\varepsilon_{n+1}}}, \ \varepsilon_{n+1} = \varepsilon'_{n+1} - i\frac{\sigma_{n+1}}{\omega}.$$

4 Synthesis Problem

The synthesis problem is reduced to an optimization problem. For solving the optimization problem we use approach similar to methods presented in [2, 3] which we outline below.

1. For the fixed ε'_j, σ_j, l_j we found $\beta_j(\omega)$. Coefficient β_j is the maximum of the reflection coefficient module for each ε'_j in region $\underline{\varepsilon} \le \varepsilon' \le \bar{\varepsilon}$, conductivity σ_j in region $\underline{\sigma} \le \sigma \le \bar{\sigma}$ and thickness l_j in region $\underline{\boldsymbol{l}} \le \boldsymbol{l} \le \bar{\boldsymbol{l}}$.

For example, for $\varepsilon_1 = 2$, $\sigma_1 = 4$, $l_1 = 1$ in region $1.8 < \omega < 10$ we have $\beta_j = 0.8$ (see Fig. 4).

2. Then we find $\eta = \min \beta_j(\varepsilon'_{\mathbf{1}}, \sigma_{\mathbf{1}}, \mathbf{l_1})$ and values of ε'_1, σ_1, l_1 corresponding to η.

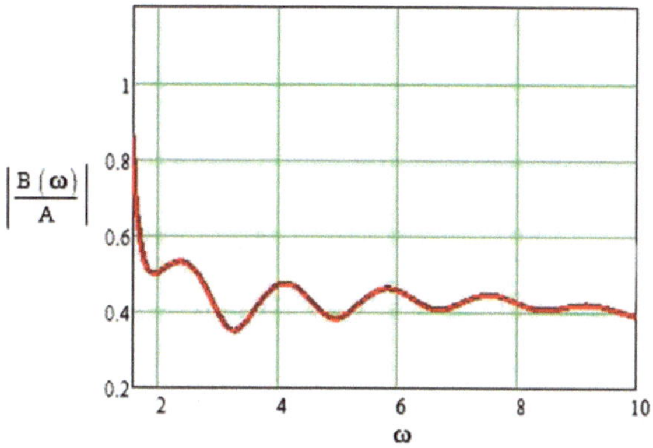

Fig. 4 High-oscillating function $|B(\omega)/A|$. Here $l_1 = 1$

5 Numerical Results

Let us consider the "perfect" conductor case.

Parameters of the two-sectional diaphragm are: $a = 2$ cm, $b = 1$ cm, the excitation frequency is $f \in (7, 63; 14, 31)$ GHz (or $\omega \in (1.6; 3)$). We have performed two set of tests for following values of parameters $\varepsilon_1{}'$, $\varepsilon_2{}'$, σ_1, σ_2, l_1, l_2:

(1) $2 \le \varepsilon_1' \le 16, 4 \le \varepsilon_2' \le 18, 0.01 \le \sigma_1 \le 0.9, 0.01 \le \sigma_2 \le 0.9, l_1 = 0.7\,\text{cm}, l_2 = 1.5$ cm.
(2) $2 \le \varepsilon_1' \le 16, 4 \le \varepsilon_2' \le 18, 0.01 \le \sigma_1 \le 0.9, 0.01 \le \sigma_2 \le 0.9, l_1 = 1\,\text{cm}, l_2 = 2$ cm.

In Table 1 we present values of $\varepsilon_1{}'$, $\varepsilon_2{}'$, σ_1, σ_2 and α. In the last column are shown values of the ratio of the minimum and maximum of the modulus of reflection coefficient.

Example 2. Parameters of the two-sectional diaphragm diaphragm are: $a = 2$ cm, $b = 1$ cm, the excitation frequency is $f \in (7, 63; 14, 31)$ GHz (or $\omega \in (1.6; 3)$).

In the Table 2 we present two different ranges for permittivity and conductivity. In the first line we show numerical results for the folowing ranges of permittivity and conductivity: $2 \le \varepsilon_1' \le 16, 4 \le \varepsilon_2' \le 18, 1.5 \le \sigma_1 \le 3, 1.5 \le \sigma_2 \le 3$. The ranges for the lengths l_1, l_2 are $0.05 \le l_1 \le 1, 0.05 \le l_2 \le 1$ cm. In the second line we present numerical results where the ranges of permittivity and conductivity are $5 \le \varepsilon_1' \le 7$, $7.5 \le \varepsilon_2' \le 9.5$, $1.5 \le \sigma_1 \le 3$, $1.5 \le \sigma_2 \le 3$, and the ranges for the lengths l_1, l_2 are $1.1 \le l_1 \le 2.1, 1 \le l_2 \le 2$.

Table 1 Example 1 for the "perfect" conductor case

$\varepsilon_1{}'$	$\varepsilon_2{}'$	σ_1	σ_2	α	ω	$\frac{min(\lvert B/A\rvert)}{max(\lvert B/A\rvert)}$
2.7	4.7	0.79	0.79	0.72	2.34	0.73
2	4	0.73	0.73	0.59	2.09	0.6

Table 2 Example 2 for the "perfect" conductor case

$\varepsilon_1{}'$	$\varepsilon_2{}'$	σ_1	σ_2	l_1	l_2	α	ω	$\frac{min(\|B/A\|)}{max(\|B/A\|)}$
10.4	12.4	0.25	0.25	0.15	0.9	0.84	3.1	0.85
6.1	8.6	0.25	0.25	1.1	2	0.59	3	0.6

Table 3 Example 3 for "good" conductor

$\varepsilon_1{}'$	$\varepsilon_2{}'$	σ_1	σ_2	l_1	l_2	α	ω	$\frac{min(\|B/A\|)}{max(\|B/A\|)}$
6.65	10.6	0.24	0.24	0.21	0.8	0.56	3.1	0.57
5	12	0.38	019	0.31	0.7	0.43	2.5	0.44

Example 3. "Good" conductor.

Parameters of the two-sectional diaphragm are: $a = 2$ cm, $b = 1$ cm, the excitation frequency is $f \in (7,63; 14,31)$ GHz (or $\omega \in (1.6; 3)$), $S = 0.038$. Let us analyze results presented in Table 3.

In the first line we show numerical results where the ranges of permittivity and conductivity are $5 \le \varepsilon_1' \le 8, 4 \le \varepsilon_2' \le 16, 1.5 \le \sigma_1 \le 3, 1.5 \le \sigma_2 \le 3$ and the range of the lengths l_1, l_2 are $0.01 \le l_1 \le 1.01, 0.05 \le l_2 \le 1$. In the second line we present numerical results where the ranges of permittivity and conductivity are $5 \le \varepsilon_1' \le 8$, $12 \le \varepsilon_2' \le 16$, $1.5 \le \sigma_1 \le 3$, $0.75 \le \sigma_2 \le 1.5$, and the ranges of the lengths l_1, l_2 are $0.01 \le l_1 \le 1.01, 0.05 \le l_2 \le 1$.

We conclude that our numerical results show validity and efficiency of the proposed technique. This technique allows construct multilayered coatings with required properties.

6 Conclusion

We developed a numerical-analytical method for the synthesis problem of a multilayered coating with minimal reflection w.r.t. complex permittivity and thickness of layers. We presented numerical results which can be applied in design of multilayered coatings.

Acknowledgements This study is supported by the Ministry of Education and Science of the Russian Federation, Project No. 1.894.2017/4.6 and by the grant of President, Project No. MK-3604.2018.1.

References

1. Y.G. Smirnov, Y.V. Shestopalov, E.D. Derevyanchuk, Permittivity reconstruction of layered dielectrics in a rectangular waveguide from the transmission coefficient at different frequencies, in *Inverse Problems and Large-Scale Computations*, vol. 52, ed. by L. Beilina, Y. Shestopalov. Series: Springer Proceedings in Mathematics and Statistics (Springer, 2013), pp. 169–181
2. Y.G. Smirnov, Y.V. Shestopalov, E.D. Derevyanchuk, Permittivity determination of multi-sectional diaphragm with metamaterial layers in rectangular waveguide, in *Proceedings of Progress in Electromagnetics Research Symposium* (Taipei, 2013), pp. 135–139
3. Y.G Smirnov, Y.V Shestopalov, E.D. Derevyanchuk, Reconstruction of permittivity and permeability tensors of anisotropic materials in a rectangular waveguide from the reflection and transmission coefficients at different frequencies, in *Proceedings of Progress in Electromagnetics Research Symposium* (Stockholm, Sweden, 2013), pp. 290–295
4. Y.G Smirnov, Y.V Shestopalov, V.V. Yakovlev, Volume singular integral equations method for determination of effective permittivity of meta- and nanomaterials, in *Proceedings of Progress in Electromagnetics Research Symposium* (Cambridge, USA, 2008), pp. 291–292
5. L. Solymar, E. Shamonina, *Waves in Metamaterials* (Oxford University Press Inc., New York, 2009)
6. D.A. Usanov, A.V. Skripal, A.V. Abramov, A.S. Bogolyubov, Determination of the metal nanometer layer thicknesss and semiconductor conductivity in metal-semiconductor structures from electromagnetic reflection and transmission spectra. Tech. Phys. **51**(5), 644–649 (2006)
7. A.V. Tikhonravov, M.K. Trubetskov, U. Keller, N. Matuschek, Designing of coatings for femtosecond lasers with phase derivatives targets, in *Advances in Optical Interference Coatings*, ed. C. Amra, A. Macleod, vol. 3738 (1999), pp. 221–229 (Proceedings of SPIE)
8. A.V. Tikhonravov, J.A. Dobrowolski, Quasi-optimal synthesis for antireflection coatings: a new method. Appl. Opt. **32**(22), 4265–4275 (1993)
9. V.B. Glasko, A.N. Tikhonov, A.V. Tikhonravov, The synthesis of multilayer coatings. USSR Comput. Math. Math. Phys. **14**(1), 135–143 (1974). https://doi.org/10.1016/0041-5553(74)90142-6

References

[illegible]

Reconstruction of Inhomogeneity Parameters by Measurements of Near Field Outside the Body

R. O. Evstigneev and M. Yu. Medvedik

Abstract In this paper we consider problems of determining the inhomogeneities inside cylindrical and rectangular bodies from the results of measuring a scalar field. These problems are reduced to the Lippmann–Schwinger equation. Observation points are located at a short distance from the body. A two-step algorithm for solving the inverse problem is used. At the first step, we reconstruct the field inside the body using the field at the observation points. At the second step, we reconstruct the inhomogeneity parameters using the obtained field values. The ill-conditioned matrix appear when using the large computational grids. In such situations, "artifacts" can appear. The presented results show the stability of approximate solutions.

Keywords Inverse problem · Cylinder · Lippmann–Schwinger equation
Parallelepiped · Numerical method

1 Introduction

The problems of determining scattered field and recovering parameters of materials arise in the applied science. A wide class of inverse problems consists of determination of certain properties of a material by measurements of the scattered field. These problems are encountered in medical diagnostics, geophysics, flaw detection, identification of explosives. One of the most important applications of this problem is the diagnosis of breast cancer. According to statistics for 2016, this disease is 23% of the total number of women with cancer. To diagnose it, X-ray mammography is used, but it is expensive and unsafe for humans. Today the method of microwave tomography [1, 2] is being developed. However, this development is constrained

R. O. Evstigneev (✉) · M. Yu. Medvedik
Penza State University, 40, Krasnaya street, Penza, Russia
e-mail: Roman_cezar@mail.ru

M. Yu. Medvedik
e-mail: _medv@mail.ru

L. Beilina and Yu. G. Smirnov (eds.), *Nonlinear and Inverse Problems in Electromagnetics*, Springer Proceedings in Mathematics & Statistics 243,
https://doi.org/10.1007/978-3-319-94060-1_3

by the lack of effective numerical methods, computational algorithms and software systems.

The direct problem is described by the Lippmann–Schwinger equation [3] and the uniqueness of the solution of this problem is considered in [4]. The inverse problem is solved using a two-step procedure. At the first step, a linear integral equation of the first kind is solved. At the second step, the geometry and parameters of the inhomogeneity are calculated using explicit formulas. It is important that there is no need to choose initial approximations that are required when applying iteration methods. In addition, the iteration methods usually require sufficiently large computations grids, in order to solve the problem with high accuracy. These advantages make it possible to consider the proposed method as very promising. The results of the calculations confirm the effectiveness of the proposed method. A number of articles [1, 2, 4, 5] are devoted to this topic.

2 Statement of the Direct Problem

We consider diffraction of a scalar wave on a body Q located in free space. The field U_0 radiated by a point source is located outside the body. This problem is solved on a cylinder in the cylindrical coordinate system. We carry out numerical experiments for two cases:

- Q is a parallelepiped,
- Q is a cylinder.

In the first case, we assume that the free space is equipped with Cartesian coordinates $Ox_1x_2x_3$. In the second case, it is equipped with cylindrical coordinates $O\rho\varphi z$ (Fig. 1).

This problem is described by the inhomogeneous equation:

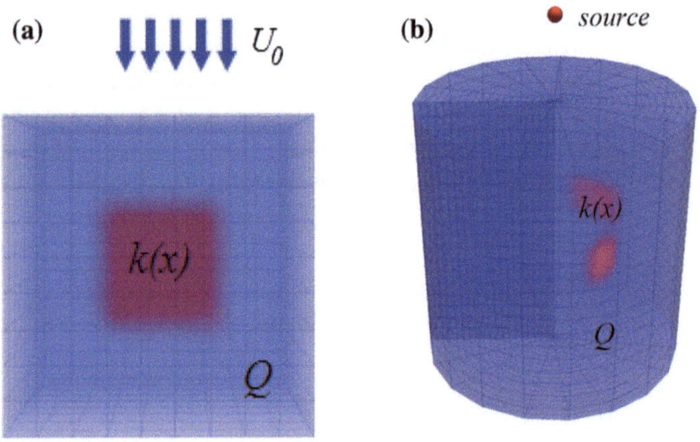

Fig. 1 The body is located in free space. The cube (**a**) and the cylinder (**b**)

$$\Delta u(x) + k^2(x)u(x) = f(x), x \in \mathbb{R}^3 \setminus (\partial Q \cup x_0), \tag{1}$$

where $k^2(x) = \begin{cases} k_1^2(x), x \in \overline{Q} \\ k_0^2, x \notin \overline{Q} \end{cases}$ is a piecewise continuous function, $f(x) = \frac{e^{ik_0|x-x_0|}}{|x-x_0|}$ and $x_0 = \left(x_0^1, x_0^2, x_0^3\right)$. We assume that on the interface between two media the transmission conditions

$$[u]_{\partial Q} = 0, \left[\frac{\partial u}{\partial n}\right]_{\partial Q} = 0 \tag{2}$$

hold. Here $[\bullet]$ is the jump of the field. In addition, the Sommerfeld radiation conditions

$$\frac{\partial u}{\partial r} = ik_0 u + o\left(\frac{1}{r}\right), \quad r := |x| \to \infty \tag{3}$$

are satisfied.

The problem (1)–(3) is reduced to the Lippmann–Schwinger equation [3]:

$$u(x) = f^0(x) + \int_Q G(x, y)\left(k_0^2 - k^2(y)\right) u(y)dy, \tag{4}$$

where $G(x, y) = \frac{e^{ik_0|x-y|}}{|x-y|}$,

$$r := |x - y| = \sqrt{(x_1 - y_1)^2 + (x_2 - y_2)^2 + (x_3 - y_3)^2}$$

in the Cartesian coordinates and

$$|x - y| = \sqrt{\rho_x^2 + \rho_y^2 - 2\rho_x\rho_y \cos(\varphi_x - \varphi_y) + (z_x - z_y)^2}$$

in the cylindrical coordinates.

We consider Eq. (4) in the space $L_2(Q)$. This is the finiteness energy condition in any bounded domain.

Denote

$$Au := \int_Q G(x, y)(k_0^2 - k^2(y)u(y)dy, \tag{5}$$

and

$$u := u(x), F := f^0(x)$$

We rewrite Eq. (4) in operator form:

$$Lu := u - Au := F, u \in L_2(Q), F \in L_2(Q), L : L_2(Q) \to L_2(Q). \tag{6}$$

The following statements hold for Eq. (6) [4]:

Statement 1 *The operator* $Lu := u - Au : L_2(Q) \to L_2(Q)$ *is a Fredholm operator with zero index.*

Statement 2 *The boundary value problem* (1)–(3) *has at most one quasi-classical solution.*

Statement 3 *The operator* $L := I - A : L_2(Q) \to L_2(Q)$ *is continuously invertible.*

Thus, the direct problem has a unique solution and is equivalent to the solution of the Lippmann–Schwinger Eq. (4).

3 Discretization of the Direct Problem

Consider a body located in free space with the Cartesian coordinate system. Let the body be a parallelepiped $Q = \{x : a_1 \le x_1 \le a_2, b_1 \le x_2 \le b_2, c_1 \le x_3 \le c_2\}$. Divide Q into elementary cells Π_{klm} in the following way (Fig. 2a):

$$\Pi_{klm} = \{x : x_k < x_1 < x_{k+1}, y_l < x_2 < y_{l+1}, z_m < x_3 < z_{m+1}\},$$
$$x_k = a_1 + \tfrac{a_2-a_1}{n}k, \; y_l = b_1 + \tfrac{b_2-b_1}{n}l, \; z_m = c_1 + \tfrac{c_2-c_1}{n}m,$$

where $k, l, m = 0, ..., n-1$. The volume of any cell Π_{klm} of the grid is V.

In the case of the cylindrical coordinates, we will consider a cylindrical body $Q = \{x : 0 \le \rho \le R, 0 \le \varphi \le 2\pi, c_1 \le z \le c_2\}$. Use the next formulas of transition to the Cartesian coordinates:

$$x = \begin{cases} x_1 = \rho\cos\varphi \\ x_2 = \rho\sin\varphi \\ x_3 = z \end{cases}$$

For the cylindrical body, the cells Π_{klm} are constructed as follows (Fig. 2b):

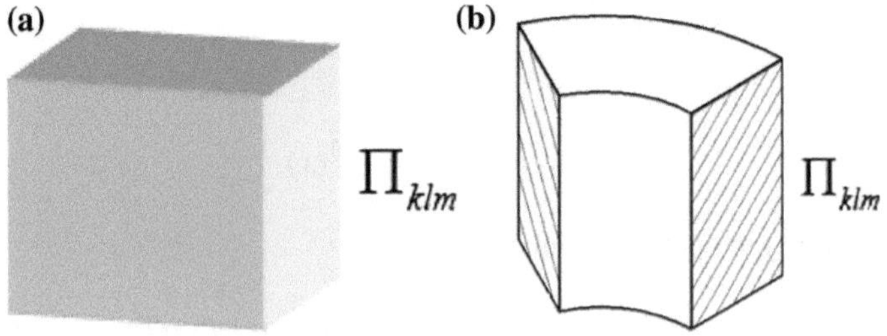

Fig. 2 Elementary cell in the cases of a parallelepiped (**a**) and a cylinder (**b**)

$$\Pi_{klm} = \{x : \rho_k < \rho < \rho_{k+1}, \varphi_l < \varphi < \varphi_{l+1}, z_m < z < z_{m+1}, \}$$
$$\rho_k = \frac{R}{n}k,\ \varphi_l = \frac{2\pi}{n}l,\ z_m = c_1 + \frac{c_2 - c_1}{n}m,$$

with $k, l, m = 0, \ldots, n-1$. The volume of each Π_{klm} is V. We introduce piecewise constant basis functions v_{klm} as follows:

$$v_{klm} = \begin{cases} 1, x \in \bar{\Pi}_{klm} \\ 0, x \notin \bar{\Pi}_{klm} \end{cases} \tag{7}$$

The set of basis functions (7) satisfies the approximation condition in $L_2(Q)$.

The collocation method for the approximate solution of equation

$$Lu = f \tag{8}$$

consists in finding an approximate solution from a finite-dimensional subspace by equating the values of the functions on the left- and the right-hand sides of the Eq. (8) at a finite number of points called collocation points [6]. Solving Eq. (8), we define the field inside the body Q.

4 Inverse Problem

The inverse problem consists in determination of the inhomogeneity inside the body from the scattered field $U(y_c)$ at the observation points y_c. Here $y = (y_1, y_2, y_3)$ is a point in the Cartesian coordinates and $y = (\rho_y, \varphi_y, z_y)$ is a point in the cylindrical coordinates. Observation points are located outside the body at short distances. The

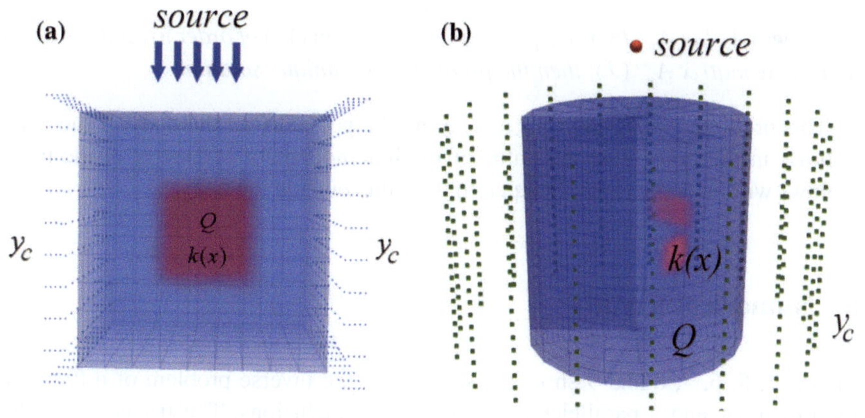

Fig. 3 The source of radiation, the body and the observation points in the cases of a parallelepiped (**a**) and a cylinder (**b**)

radiation source is located at a certain distance from the body and radiated a constant field U_0. Suppose that Q contains one or more inhomogeneities. We divide the body into elementary cells Π_{klm} and assume that inside each cell $k(x) = k_i$, where k_i are constants (Fig. 3).

To find the wave number $k(x)$ inside a cylindrical body Q, we use the following algorithm [7].

At the first step, we solve the integral equation:

$$U(y_c) = \int\limits_Q G(x, y_c)J(x)dx + f(y_c). \tag{9}$$

We reconstruct the field $J(y)$, where $U(y_c)$ is the known measured filed. At the second step, we substitute the previously obtained $J(y)$ into the expression

$$\frac{J(y)}{k^2(y) - k_0^2} - \int\limits_Q G(x, y)\, J(x)\, dx = f(y) \tag{10}$$

and determine $k(y)$.

We introduce the following notation:

$$J_n(x) = \sum_{k,l,m=0}^{n-1} c_{klm} v_{klm}$$

and

$$\int\limits_Q G(x, y_c^j) J_n(x) dx = U(y_c^j) - f(y_c^j),\ y_c^j \notin \bar{Q},\ j = 0, \ldots, n^3 - 1. \tag{11}$$

Statement 4 *Let* $A_n(J)$ *be a square matrix of system* (11) *of order* n^3. *If there exists an inverse matrix* $A_n^{-1}(J)$, *then the problem has a unique solution.*

This problem can be used for recovering the geometry of the body. If there are hollows in the body, then the value of the inhomogeneity parameter inside these hollows will be approximately equal to the value outside the body.

5 Numerical Results

Figures 4, 5, 6, 7, 8 and 9 show the solution of the inverse problem of diffraction by a cylinder and a parallelepiped with several inclusions. The frequency of the wave is 2.0 GHz. The results presented below show the reconstruction of the real

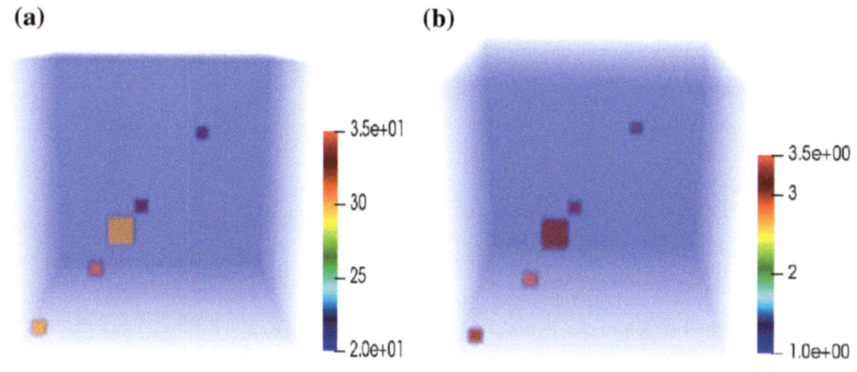

Fig. 4 The original structure of a rectangular body: the real part (**a**) and the imaginary part (**b**)

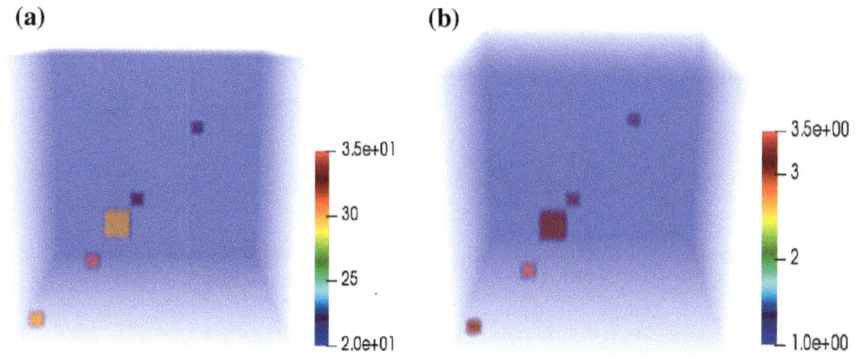

Fig. 5 The reconstructed structure of a rectangular body: the real part (**a**) and the imaginary part (**b**)

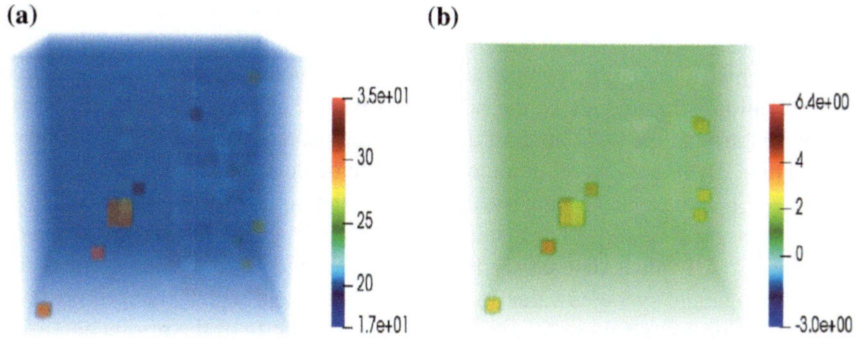

Fig. 6 Reconstructing the structure of a rectangular body with measurement errors: the real part (**a**) and the image part (**b**)

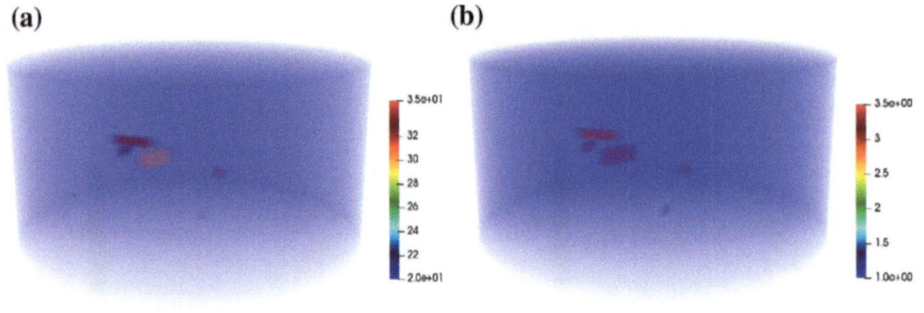

Fig. 7 The original structure of a cylindrical body: the real part (**a**) and the imaginary part (**b**)

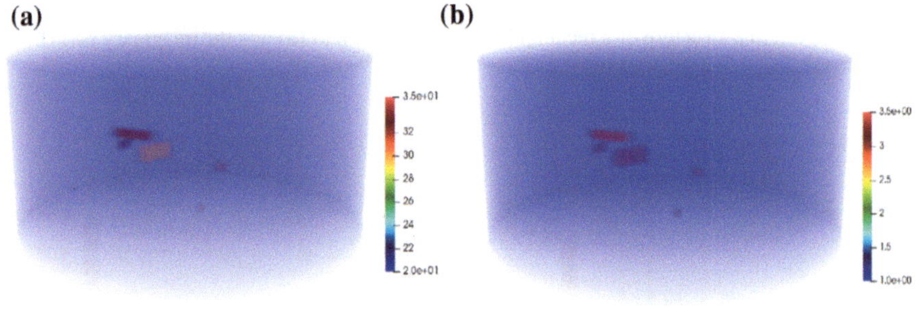

Fig. 8 The reconstructed structure of a cylindrical body: the real part (**a**) and the imaginary part (**b**)

and imaginary parts of the wave function. The radius and the height of the cylinder equal 7.5 cm. The length of each side of the rectangular body is 15 cm. The partition numbers on a fine grid are $k = l = m = 20$, whereas on a coarse grid we use $k = l = m = 4$, which allowed reducing the value of the matrix condition to the acceptable value $cond(L) < 10^8$. For such condition numbers, the solution is sufficiently stable to errors, which allows us to reconstruct the internal structure of the body. The source of observation is located above the body at a distance of 3 cm. Observation points are located around the body at short distances, the nearest ones are at a distance of about 1 mm. There are several inhomogeneities to be detected.

The Fig. 6 represent the numerical results of the solution with an error in U_c. To obtain a good quality of the solution, "artifacts" are eliminated by repeated experiments. All values that satisfy the allowable range of background values are equated to a single averaged value. A repeated experiment with a changed position of the radiation source allows one to verify the authenticity of the results obtained, as well as to get rid of the "artifacts".

The same results are obtained for a cylindrical body. This body shape makes it possible to arrange observation points more evenly, which allows obtaining more favorable results and improving the quality of the solutions. Repeated experiments on this configuration can be carried out by shifting the position of the observation points by 1°.

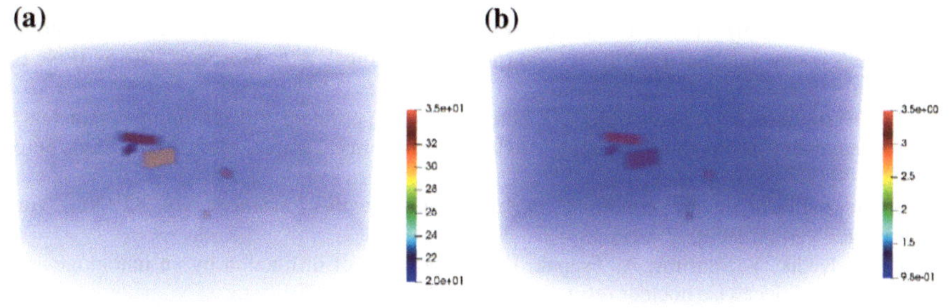

Fig. 9 The reconstructed structure of a cylindrical body with measurement errors: the real part (**a**) and imaginary part (**b**)

6 Conclusion

The presented results show that the algorithm used is applicable in the case of "noisy data". The maximum modulus of error does not exceed 7% of the exact value. The numerical algorithm is stable to measurement errors and has the following advantages over the iterative algorithms [2–6]:

- using the two-step method;
- the possibility of refining the solution at the points of observation;
- the initial approximation is not required for calculations;
- the ability to filter out artifacts by solving the additional problems.

7 Funding

The reported study was funded by RFBR according to the research project 18-01-00219 and the research project 18-31-00108.

References

1. P.M. Meaney, Q. Fang, T. Rubaek, E. Demidenko, and K.D. Paulsen, Log transformation benefits parameter estimation in microwave tomographic imaging. Med. Phys. **34**(6), 2014–2023 (2007)
2. T.M. Grzegorczyc, P.M. Meaney, P.A. Kaufman, Roberta M. diFlorio-Alexander, K.D. Paulsen, Fast 3-D tomographic microwave imaging for breast cancer detection. IEEE Trans. Med. Imaging. **3**(8), 1584–1592 (2012)
3. M.Y. Medvedik, A subhierarhic method for solving the Lippmann–Schwinger integral equation on bodies of complex shapes. J. Comput. Technol. Electron. **57**(2), 158–163 (2012)

4. Y.G. Smirnov (2016), Diffraction of acoustic and electromagnetic waves by screens and inhomogeneous solids: mathematical theory/Smirnov Yu.G., Tsupak A.A. Moscow: "RU-SCIENCE", p. 214
5. R.O. Evstigneev, M.Y. Medvedik, Y.G. Smirnov, Inverse problem of determining parameters of inhomogeneity of a body from acoustic field measurements. Comput. Math. Math. Phys. **56**(3), 483-490 (2016)
6. R. Kress, Linear integral equations, in *Applied Mathematical Sciences*, vol. 82 (New York, Springer-Verlag, 1989)
7. M. Medvedik, Y. Smirnov, A. Tsupak, Inverse problem of diffraction by an inhomogeneous solid with a piecewise hoelder refractive index arXiv:1803.04701

Diffraction of TE Polarized Electromagnetic Waves by a Layer with a Nonlinear Medium

Yu. G. Smirnov, E. Smolkin and V. Kurseeva

Abstract The diffraction of a plane wave by a dielectric layer filled with a nonlinear medium is considered. The layer is located between two half-spaces with constant permittivities. Two widely used types of nonlinearities (Kerr nonlinearity and nonlinearity with saturation) are considered. The problem is to find amplitudes of reflected and transmitted fields. An amplitude of the incident field is supposed to be known. The analytical and numerical solution techniques are developed. Numerical results are presented.

Keywords Maxwell's equations · Kerr nonlinearity · Saturated nonlinearity

1 Introduction

The electromagnetic wave diffraction by homogeneous (see [1] where early and classical results for circular geometries are summarized) or inhomogeneous (see e.g. [2]) dielectric layers filled with linear medium has been a subject of intense studies since the late 1940s. However, the case of nonlinear inhomogeneous filling still constitutes to a big extent an open problem, both in view of mathematical justification and creation of efficient numerical techniques.

However, the case of nonlinear filling still constitutes to a big extent an open problem, both in view of mathematical justification and creation of efficient numerical techniques. A progress here is associated with recently developed techniques [3–5] for the analysis of nonlinear boundary value problems for the Maxwell and Helmholtz equations.

Yu. G. Smirnov · E. Smolkin · V. Kurseeva (✉)
Penza State University, 40, Krasnaya street, Penza, Russia
e-mail: 79273698109@ya.ru

Yu. G. Smirnov
e-mail: smirnovyug@mail.ru

E. Smolkin
e-mail: smolkin@hotmail.com

L. Beilina and Yu. G. Smirnov (eds.), *Nonlinear and Inverse Problems in Electromagnetics*, Springer Proceedings in Mathematics & Statistics 243,
https://doi.org/10.1007/978-3-319-94060-1_4

The objectives of this study are to fill these gaps in the electromagnetic theory and numerical methods aimed at efficient solution to the problems of diffraction by layers filled with nonlinear media.

2 Statement of the Problem

Let us consider a monochromatic TE wave $(\mathbf{E}, \mathbf{H})e^{-i\omega t}$, where ω is the circular frequency and

$$\mathbf{E} = (0, E_y, 0)^\top, \quad \mathbf{H} = (H_x, 0, H_z)^\top \tag{1}$$

are the complex amplitudes [6]. The TE wave diffract by the lossless dielectric layer $\Sigma = \{(x, y, z) : 0 \leqslant x \leqslant h, (y, z) \in \mathbb{R}^2\}$, where $h > 0$. The space is filled with a homogeneous isotropic nonmagnetic media. The permittivity in the entire space has the form $\varepsilon = \widetilde{\varepsilon}\varepsilon_0$, where

$$\widetilde{\varepsilon} = \begin{cases} \varepsilon_1, & x > h, \\ \varepsilon_2 + \widetilde{\alpha} f(|\mathbf{E}|^2), & 0 \leqslant x \leqslant h, \\ \varepsilon_3, & x < 0, \end{cases}$$

$\varepsilon_1, \varepsilon_2, \varepsilon_3, \widetilde{\alpha} > 0$ are real constants and $\varepsilon_0 > 0$ is the permittivity of vacuum. We assume that $\varepsilon_2 > \max\{\varepsilon_1, \varepsilon_3\}$ and $\min\{\varepsilon_1, \varepsilon_3\} \geqslant \varepsilon_0$.

We also assume that the components of complex amplitudes (1) have the form

$$E_y = \mathrm{E}_y(x)e^{i\gamma z}, \quad H_x = \mathrm{H}_x(x)e^{i\gamma z}, \quad H_z = \mathrm{H}_z(x)e^{i\gamma z}, \tag{2}$$

where γ is a known real constant, E_y, H_x, H_z are unknown functions.

Complex amplitudes (1) exponentially decay as $|x| \to \infty$ in the domains $x < 0$ and $x > h$.

In the half-space $x \geqslant h$ complex amplitudes (1) satisfy Maxwell's equations

$$\begin{cases} \operatorname{rot} \mathbf{H} = -i\omega\varepsilon_0\varepsilon_1\mathbf{E} + \mathbf{j}^{\text{ext}}, \\ \operatorname{rot} \mathbf{E} = i\omega\mu_0\mathbf{H}, \end{cases} \tag{3}$$

where $\mathbf{j}^{\text{ext}} = \mathbf{e}_y iAe^{i\gamma z}\delta(x - x_0)$, μ_0 is the permeability of vacuum, x_0 is a real constant.

Substituting complex amplitudes (1) with components (2) into Maxwell's equations (3), we obtain

$$\begin{cases} i\gamma\mathrm{H}_x - \mathrm{H}_z' = -i\omega\varepsilon_0\varepsilon_1\mathrm{E}_y + iA\delta(x - x_0), \\ i\gamma\mathrm{E}_y = -i\omega\mu_0\mathrm{H}_x, \\ \mathrm{E}_y' = i\omega\mu_0\mathrm{H}_z, \end{cases} \tag{4}$$

where $(\cdot)' \equiv \partial_x$.

From system (4), we obtain

$$H_x = -\frac{\gamma}{\omega\mu_0}E_y, \quad H_z = -\frac{i}{\omega\mu_0}E_y'. \tag{5}$$

Using (5), we find

$$E_y'' + (k_0^2\varepsilon_1 - \gamma^2)E_y = A\omega\mu_0\delta(x - x_0), \quad x \geqslant h, \tag{6}$$

where $k_0^2 = \omega^2\mu_0\varepsilon_0$.

In accordance with the radiation condition, solution of Eq. (6) is

$$E_y(x) = C_1 e^{-\kappa_1(x-h)} - \frac{A\omega\mu_0 e^{-\kappa_1|x-x_0|}}{2\kappa_1}, \quad x \geqslant h,$$

where $\kappa_1^2 = \gamma^2 - k_0^2\varepsilon_1 > 0$. Thus the tangential component of the electric field and its derivative at the point $x = h$ have the form

$$E_y(h) = F_R - F_I, \quad E_y'(h) = -\kappa_1(F_R + F_I), \tag{7}$$

where $F_R = C_1$ is an amplitude of the reflected field, $F_I = \frac{A\omega\mu_0 e^{-\kappa_1(x_0-h)}}{2\kappa_1}$ is an amplitude of the incident field at the point $x = h$.

In the layer Σ and in the half-space $x \leqslant 0$ complex amplitudes (1) satisfy Maxwell's equations

$$\begin{cases} \operatorname{rot}\mathbf{H} = -i\omega\varepsilon\mathbf{E}, \\ \operatorname{rot}\mathbf{E} = i\omega\mu_0\mathbf{H}, \end{cases} \tag{8}$$

Substituting complex amplitudes (1) with components (2) into Maxwell's equations (8), we obtain

$$\begin{cases} i\gamma H_x - H_z' = -i\omega\varepsilon E_y, \\ i\gamma E_y = -i\omega\mu_0 H_x, \\ E_y' = i\omega\mu_0 H_z. \end{cases} \tag{9}$$

From system (9), we obtain

$$H_x = -\frac{\gamma}{\omega\mu_0}E_y, \quad H_z = -\frac{i}{\omega\mu_0}E_y'. \tag{10}$$

Using (10), we find

$$E_y'' - \gamma^2 E_y = -\omega^2\mu_0\varepsilon E_y. \tag{11}$$

In the half-space $x \leqslant 0$ Eq. (11) has the form

$$\mathrm{E}_y'' = \kappa_3^2 \mathrm{E}_y, \tag{12}$$

where $\kappa_3^2 = \gamma^2 - k_0^2 \varepsilon_3 > 0$.

In accordance with the radiation condition, solution of Eq. (12) is

$$\mathrm{E}_y(x) = C_3 e^{\kappa_3 x}.$$

Thus the tangential component of the electric field and its derivative at the point $x = 0$ have the form

$$\mathrm{E}_y(0) = F_T, \quad \mathrm{E}_y'(0) = \kappa_3 F_T, \tag{13}$$

where $F_T = C_3$ is an amplitude of the transmitted field.

Using the notation $u := E_y$, in the layer Σ Eq. (11) takes the form

$$u'' = -(\kappa_2^2 - \alpha f(u^2))u. \tag{14}$$

where $\kappa_2^2 = k_0^2 \varepsilon_2 - \gamma^2, \alpha = k_0^2 \widetilde{\alpha}$.

Since E_y and H_z are continuous at the interfaces $x = 0$, $x = h$, then one has the following transmission conditions

$$[u]|_{x=0} = 0, \; [u']\big|_{x=0} = 0, \; [u]|_{x=h} = 0, \; [u']\big|_{x=h} = 0, \tag{15}$$

where $[v]|_{x=x_0} = \lim_{x \to x_0 - 0} v(x) - \lim_{x \to x_0 + 0} v(x)$. It follows from (15) that

$$u(0) = F_T, \qquad u'(0) = \kappa_3 F_T, \tag{16}$$

$$u(h) = F_R - F_I, \qquad u'(h) = -\kappa_1 (F_R + F_I). \tag{17}$$

Diffraction problem P: *The goal is to find the nontrivial function* $u(x) \in C^1(-\infty, +\infty) \cap C^2(-\infty, 0) \cap C^2(0, h) \cap C^2(h, +\infty)$ *that is defined by formula*

$$u(x) = F_T e^{\kappa_3 x},$$

for $x < 0$, *is defined by formula*

$$u(x) = F_R e^{-\kappa_1 (x-h)} - F_I e^{-\kappa_1 \left(|x - x_0| + (h - x_0)\right)},$$

for $x > h$ *and solves equation* (14) *for* $0 \leqslant x \leqslant h$; *moreover, the function* $u(x)$ *thus defined satisfies transmission conditions* (15).

Using the function $u(x)$, the amplitude of the reflected F_R and transmitted F_T field can be determined for a given value of the amplitude of the incident field F_I. This is of great importance for practical application.

Remark 1. If the function u is a solution of the problem P, then the function $-u$ is also a solution of the problem P. For this reason it is enough to study the case $F_T > 0$.

3 Saturated Nonlinearity

Let us consider the case of saturable nonlinear permittivity. We consider the permittivity of the form $f(u^2) = \frac{u^2}{1+\beta u^2}$, where $\beta > 0$ is a real constant.

Let $L = \frac{d^2}{dx^2} + \kappa_2^2$. Then Eq. (14) can be rewritten as

$$Lu = -\frac{\alpha u^3}{1+\beta u^2}.$$

Below we need to consider the following boundary value problem

$$\begin{cases} LG = -\delta(x-s), \\ G'\big|_{x=0} = G'\big|_{x=h} = 0. \end{cases} \tag{18}$$

It can be proved that the Green function has the form:

$$G(x,s) = \begin{cases} -\frac{\cos(\kappa_2 x)\cos(\kappa_2(s-h))}{\kappa_2 \sin(\kappa_2 h)}, & x < s \leqslant h, \\ -\frac{\cos(\kappa_2 (x-h))\cos(\kappa_2 s)}{\kappa_2 \sin(\kappa_2 h)}, & s < x \leqslant h. \end{cases}$$

Using the second Green's formula, we obtain

$$\int_0^h (GLu - uLG)\,dx = \left(u'(x)G(x,s) - u(x)G'(x,s)\right)\big|_0^h.$$

Using Eqs. (7), (13) and (18), from the last equation, we find an integral representation of the solution $u(s)$ of Eq. (14) at $s \in [0, h]$:

$$u(s) = \int_0^h G(x,s)\frac{\alpha u^3(x)}{1+\beta u^2(x)}dx - (F_R + F_I)\frac{\kappa_1 \cos(\kappa_2 s)}{\kappa_2 \sin(\kappa_2 h)} - \kappa_3 F_T \frac{\cos(\kappa_2(s-h))}{\kappa_2 \sin(\kappa_2 h)}. \tag{19}$$

Setting $s = h$ in (19), we have

$$F_R = \frac{F_I D_I}{D_R} - \frac{\kappa_3 F_T}{\kappa_2 \sin(\kappa_2 h) D_R} - \frac{\alpha}{D_R}\int_0^h \frac{\cos(\kappa_2 x)}{\kappa_2 \sin(\kappa_2 h)}\frac{u^3(x)}{1+\beta u^2(x)}dx, \tag{20}$$

where

$$D_R = 1 + \frac{\kappa_1}{\kappa_2}\cot(\kappa_2 h),$$

$$D_I = 1 - \frac{\kappa_1}{\kappa_2}\cot(\kappa_2 h).$$

Setting $s = 0$ in (19), we have

$$F_T = -\frac{\kappa_1(F_R + F_I)}{\kappa_2 \sin(\kappa_2 h) D_T} - \frac{\alpha}{D_T}\int_0^h \frac{\cos(\kappa_2(x-h))}{\kappa_2 \sin(k_2 h)} \frac{u^3(x)}{1+\beta u^2(x)} dx, \tag{21}$$

where

$$D_T = 1 + \frac{\kappa_3}{\kappa_2}\cot(\kappa_2 h).$$

Substituting (20) and (21) into Eq. (19), we obtain

$$u(s) = \alpha \int_0^h Q_1(x,s) \frac{u^3(x)}{1+\beta u^2(x)} dx + Q_2(s), \tag{22}$$

where

$$Q_1(x,s) = G(x,s) + \frac{\kappa_1 \cos(\kappa_2 s)\cos(\kappa_2 x)}{\kappa_2^2 D_R \sin^2(k_2 h)} + \frac{\kappa_3 \cos(\kappa_2(s-h))\cos(\kappa_2(x-h))}{\kappa_2^2 D_T \sin^2(k_2 h)},$$

$$Q_2(s) = \frac{\kappa_1\kappa_3 \cos(\kappa_2 s) F_T}{\kappa_2^2 D_R \sin^2(k_2 h)} - \frac{\kappa_1 \cos(\kappa_2 s) F_I (D_I + D_R)}{\kappa_2 D_R \sin(k_2 h)} + \\ + \frac{\kappa_1\kappa_3 \cos(\kappa_2(s-h))(F_R + F_I)}{\kappa_2^2 D_T \sin(\kappa_2 h)^2}.$$

Statement 1 *Let* $\alpha > 0$, $\beta > 0$, *and* $q = \frac{3\alpha}{2\beta} M < 1$, *where*

$$M = \max_{0 \leqslant s \leqslant h} \int_0^h |Q_1(x,s)|\, dx.$$

Then Eq. (22) *has a unique solution* $u = u^* \in C[0,h]$.

Proof. It is easy to check that the following estimate

$$\left| \frac{u_1^3}{1+\beta u_1^2} - \frac{u_2^3}{1+\beta u_2^2} \right| \leqslant \frac{3}{2\beta}|u_1 - u_2| \tag{23}$$

is valid. Let us rewrite Eq. (22) in the operator form

$$u = \alpha \mathrm{Q}_1 u + \mathrm{Q}_2; \quad \mathrm{Q}_1 : C[0, h] \to C[0, h]. \tag{24}$$

In accordance with estimate (23), we obtain

$$|\alpha \mathrm{Q}_1 u_1 - \alpha \mathrm{Q}_1 u_2| \leqslant \frac{3\alpha}{2\beta} \max_{0 \leqslant s \leqslant h} |u_1 - u_2| \int_0^h |Q_1(x, s)|\, dx$$

and hence

$$\|\alpha \mathrm{Q}_1 u_1 - \alpha \mathrm{Q}_1 u_2\| \leqslant \frac{3\alpha}{2\beta} M \|u_1 - u_2\|,$$

where $M = \max_{0 \leqslant s \leqslant h} \int_0^h |Q_1(x, s)|\, dx$ and $\|u\| = \max_{0 \leqslant s \leqslant h} |u|$. Clearly, for $q = \frac{3\alpha}{2\beta} M < 1$ the mapping $\alpha \mathrm{Q}_1$ is a contraction mapping. According to Corollary 2 (see [7] p. 392) the operator Q_1 maps the ball $B_r(0)$ in the same ball in $C[0, h]$. We choose $r > 0$ in such a way that the condition $\|\mathrm{Q}_2\| \leqslant (1 - q)r$ is fulfilled. Then Eq. (24) has a unique solution $u^* \in B_r(0) \subset C[0, h]$. Since r can be chosen arbitrarily large, then the statement is valid.

4 Kerr Nonlinearity

Let us consider the case in which permittivity is described by the Kerr law. In this case $f(u^2) = u^2$. The first integral of Eq. (14) has a form

$$u'^2 = -\kappa_2^2 u^2 - \tfrac{1}{2}\alpha u^4 + C.$$

where C is a real constant.

Using transmission conditions (7), (13), we obtain

$$\begin{cases} C = (\varepsilon_2 - \varepsilon_3) k_0^2 F_T^2 + \frac{1}{2}\alpha F_T^4, \\ C = \kappa_1^2 (F_R + F_I)^2 + \kappa_2^2 (F_R - F_I)^2 + \frac{1}{2}\alpha (F_R - F_I)^4, \end{cases} \tag{25}$$

From system (25) we finds

$$F_T^2 = \frac{(\varepsilon_3 - \varepsilon_2) k_0^2 + \sqrt{(\varepsilon_2 - \varepsilon_3)^2 k_0^4 + 2\alpha C}}{\alpha},$$

where

$$C = \kappa_1^2 (F_R + F_I)^2 + \kappa_2^2 (F_R - F_I)^2 + \tfrac{1}{2}\alpha (F_R - F_I)^4. \tag{26}$$

In accordance with Remark 1, we obtain

$$F_T = \sqrt{\frac{(\varepsilon_3 - \varepsilon_2)k_0^2 + \sqrt{(\varepsilon_2 - \varepsilon_3)^2 k_0^4 + 2\alpha C}}{\alpha}}, \tag{27}$$

where C is given by (26)

By introducing new variables

$$\tau(x) := u^2(x), \quad \eta(x) := \frac{u'(x)}{u(x)},$$

Equation (14) can be rewritten as

$$\begin{cases} \tau' = 2\tau\eta, \\ \eta' = -\left(\eta^2 + \kappa_2^2 + \alpha\tau\right) \end{cases} \tag{28}$$

with the first integral

$$\tfrac{1}{2}\alpha\tau^2 + \left(\eta^2 + \kappa_2^2\right)\tau = C. \tag{29}$$

Solving (29) with respect to τ, taking into account that $\tau \geqslant 0$, and substituting the result into the right-hand side of the second equation in (28), one obtains

$$\eta' = -\sqrt{\left(\eta^2 + \kappa_2^2\right)^2 + 2\alpha C}, \tag{30}$$

where the radicand is positive for all real η and γ and $\alpha C > 0$.

Using boundary conditions (16) and (17), we find

$$\eta(0) = -\kappa_3, \quad \eta(h) = -\kappa_1 \frac{F_R + F_I}{F_R - F_I}. \tag{31}$$

Since $\eta' < 0$, then η monotonically decreases for $x \in (0, h)$. However, η is continuous if and only if $u(x)$ does not vanish for all $x \in (0, h)$. In the general case, $u(x)$ can have zeros at some points on the interval $(0, h)$. Suppose that $u(x)$ has n zeros $x_1, \ldots, x_n \in (0, h)$. Then η has n break points $x_1, \ldots, x_n \in (0, h)$. It is clear that $E_y'(x_i) \neq 0$ for all $i = \overline{1, n}$. Using formula (30), we obtain

$$\eta(x_i - 0) = -\infty, \quad \eta(x_i + 0) = +\infty, \quad i = \overline{1, n}. \tag{32}$$

Introduce the notation

$$w := \frac{1}{\sqrt{\left(s^2 + \kappa_2^2\right)^2 + 2\alpha C}}, \tag{33}$$

where $w \equiv w(s; C)$.

The solution of Eq. (30) will be sought in the intervals $[0, x_1)$, (x_1, x_2), ..., $(x_{n_1}, h]$:

$$\begin{cases} \int\limits_{\eta(x)}^{\eta(x_1-0)} w ds = x + c_0, \quad 0 \leqslant x < x_1; \\ -\int\limits_{\eta(x_i+0)}^{\eta(x)} w ds = x + c_i, \quad x_i < x < x_{i+1}, \quad i = \overline{0, n-1}; \\ -\int\limits_{\eta(x_n+0)}^{\eta(x)} w ds = x + c_n, \quad x_n < x \leqslant h. \end{cases} \tag{34}$$

Substituting $x = 0$, $x = x_{i+1} - 0$, $x = h$ in (34) (in the first, second and third formulas, respectively), we find the constants $c_0, c_1, \ldots, c_n$:

$$\begin{cases} c_0 = \int\limits_{\eta(0)}^{\eta(x_1-0)} w ds; \\ c_i = -\int\limits_{\eta(x_i+0)}^{\eta(x_{i+1}-0)} w ds - x_{i+1}, \quad i = \overline{1, n-1}; \\ c_n = -\int\limits_{\eta(x_n+0)}^{\eta(h)} w ds - h. \end{cases} \tag{35}$$

Taking into account (35), we rewrite (34) in the following form

$$\begin{cases} \int\limits_{\eta(x)}^{\eta(x_1-0)} w ds = x + \int\limits_{\eta(0)}^{\eta(x_1-0)} w ds, & 0 \leqslant x < x_1; \\ -\int\limits_{\eta(x_i+0)}^{\eta(x)} w ds = x - \int\limits_{\eta(x_i+0)}^{\eta(x_{i+1}-0)} w ds - x_{i+1}, & x_i < x < x_{i+1}; \\ -\int\limits_{\eta(x_n+0)}^{\eta(x)} w ds = x - \int\limits_{\eta(x_n+0)}^{\eta(h)} w ds - h, & x_n < x \leqslant h. \end{cases} \tag{36}$$

Substituting $x = x_1 - 0$, $x = x_i + 0$, $x = x_n + 0$ in (36) (in the first, second and third formulas respectively), we obtain

$$\begin{cases} 0 = x_1 + \int\limits_{\eta(0)}^{\eta(x_1-0)} w ds, \\ 0 = x_i - \int\limits_{\eta(x_i+0)}^{\eta(x_{i+1}-0)} w ds - x_{i+1}, \quad i = \overline{1, n-1}; \\ 0 = x_n - \int\limits_{\eta(x_n+0)}^{\eta(h)} w ds - h. \end{cases} \tag{37}$$

Taking into account (31), (32) and using (37), we obtain

$$\begin{cases} 0 < x_1 = \int\limits_{-\kappa_3}^{+\infty} w ds, \\ 0 < x_{i+1} - x_i = \int\limits_{-\infty}^{+\infty} w ds, \quad i = \overline{1, n-1}; \\ 0 < h - x_n = \int\limits_{\eta(h)}^{+\infty} w ds. \end{cases} \tag{38}$$

Formulas (38) give explicit expressions for distances between zeros of u. Moreover, since the left-hand sides in (38) are finite, the right-hand sides are also finite. Therefore, the improper integrals on the right-hand sides converge.

Summing (38), we obtain

$$x_1 + x_2 - x_1 + x_3 - x_2 + \ldots + x_{n-1} - x_{n-2} + x_n - x_{n-1} + h - x_n =$$
$$= \int\limits_{-\kappa_3}^{+\infty} w ds + (n-1) \int\limits_{-\infty}^{+\infty} w ds + \int\limits_{\eta(h)}^{+\infty} w ds.$$

From the last equation, we obtain

$$\int\limits_{-\kappa_3}^{-\kappa_1 \frac{F_R+F_I}{F_R-F_I}} w ds + (n+1) \int\limits_{-\infty}^{+\infty} w ds = h, \tag{39}$$

where $n = 0, 1, 2, \ldots$

Formula (39) shows that the solution of the Cauchy problem to Eq. (14) with initial conditions (16) exists and is defined globally at any segment $[0, h]$. The uniqueness of this solution and its continuity with respect to γ follows from smoothness of the right-hand side of Eq. (14) with respect to u and γ [8].

It is more convenient to exclude the variable C from Eq. (33). This can be done using (26):

$$w := \frac{1}{\sqrt{\left(s^2 + \kappa_2^2\right)^2 + 2\alpha\left((\varepsilon_2 - \varepsilon_3)k_0^2 F_T^2 + \frac{1}{2}\alpha F_T^4\right)}},$$

where $n = 0, 1, 2, \ldots$ From the arguments given above yields the following.

Statement 2 *Let $\alpha > 0$, h and F_I are known. For some integer $n \geqslant 0$ there is F_R such that F_R and F_I satisfy Eq. (39).*

Statement 3 *The Cauchy problem for Eq. (14) with initial conditions (16), where $F_T > 0$ and $\kappa_3 = \sqrt{\gamma^2 - k_0^2\varepsilon_3} > 0$ are constants, has a unique continuous solution $u \equiv u(x; \gamma)$ defined globally on $[0, x^*]$, where $x^* > 0$ is an arbitrary real point. This solution depends continuously on γ for all $\gamma^2 > k_0^2\varepsilon_3$.*

Theorem 1. *Let $\alpha > 0$, h and F_I are known. Problem P has an infinite number of solutions in the Kerr case.*

Proof. It follows from the Statement 2 that for given $\alpha > 0$, h and F_I there exists an infinite number of values $F_R^{(n)}$ such that $F_R^{(n)}$ and F_I satisfy the Eq. (39) for some integer $n \geqslant 0$. Let us prove that if $n \neq m$ then $F_R^{(n)} \neq F_R^{(m)}$. Let $n \neq m$, but $F_R^{(n)} = F_R^{(m)}$. Let us consider the following

$$\int_{-\kappa_3}^{-\kappa_1 \frac{F_R^{(n)}+F_I}{F_R^{(n)}-F_I}} w ds + (n+1) \int_{-\infty}^{+\infty} w ds - \int_{-\kappa_3}^{-\kappa_1 \frac{F_R^{(m)}+F_I}{F_R^{(m)}-F_I}} w ds - (m+1) \int_{-\infty}^{+\infty} w ds = 0.$$

From the last expression, we obtain

$$(n-m) \int_{-\infty}^{+\infty} w ds = 0.$$

Since $w > 0$, then $n = m$.

Each pair $F_R = F_R^{(n)}$ and F_I determines value $F_R = F_T^{(n)}$ in accordance with Eq. (27). It follows from Statement 3 [5] that the Cauchy problem for Eq. (14) with initial conditions

$$u(0) = F_T^{(n)}, \ u'(0) = \kappa_3 F_T^{(n)}$$

has a global unique continuous solution for each value $F_T^{(n)}$. Therefore problem P has an infinite number of solutions.

5 Numerical Results

For numerical solution of Problem P a method based on the solution to the auxiliary Cauchy problem is proposed [9] which makes it possible, in particular, to determine and plot the amplitude F_R of the reflected field, with respect to the amplitude F_I of the incident field.

The following parameters are used for calculations: $h = 2$ mm, $x_0 = 3$ mm, $\omega = 1$ GHz, $\mu_0 = 1$, $\varepsilon_1 = 1$, $\varepsilon_2 = 2$, $\varepsilon_3 = 1$, $\gamma = 1.1$ mm^{-1}, $\alpha = 0.01$ V^{-2}.

We have carried out experiments for two widely used types of nonlinearities. In each experiment we compared the nonlinear and the linear cases.

In the first experiment we chose $f(u^2) = u^2$ (Figs. 1, 2, 3 and 4). In the second experiment we chose $f(u^2) = \frac{u^2}{1+\beta u^2}$, with $\beta = 0.005$ V^{-2} (Figs. 5 and 6).

In Figs. 1, 5 the amplitude F_R of the reflected field calculated with respect to the amplitude F_I of the incident field are shown. Red lines correspond to the linear case, blue curves correspond to the nonlinear cases. The vertical dashed black lines

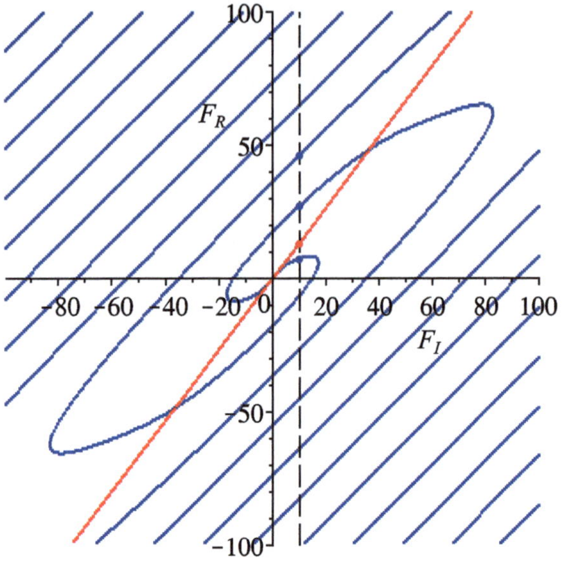

Fig. 1 The graph of curves for the Kerr case. Marked points of the amplitudes of the reflected field: for the linear case $F_R \approx 13.325$ (red dot); for the Kerr's case $F_R \approx 7.642$, $F_R \approx 27.337$, $F_R \approx 46.479$ (blue dots)

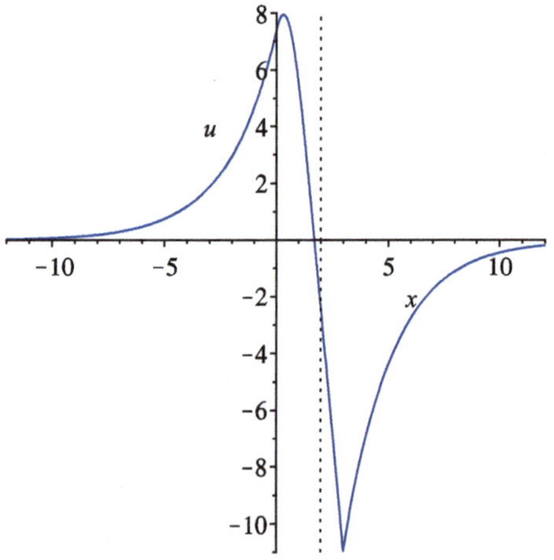

Fig. 2 The graph of u for the Kerr case. $F_R \approx 7.642$

correspond to the $F_I = 10$. Points of intersections of the vertical dashed lines with the curves are the amplitudes of the reflected field.

For the value of the amplitude of the incident field $F_I = 10$ the graphs of the solution u in Figs. 2, 3, 4, 6 and 7 are presented. The vertical grey dashed lines correspond to the boundary $x = h$. Graphs of the tangential components of the electromagnetic field consistent with the physical formulation of the problem, namely solutions u are continuous at the interface and decay when $x \to \infty$.

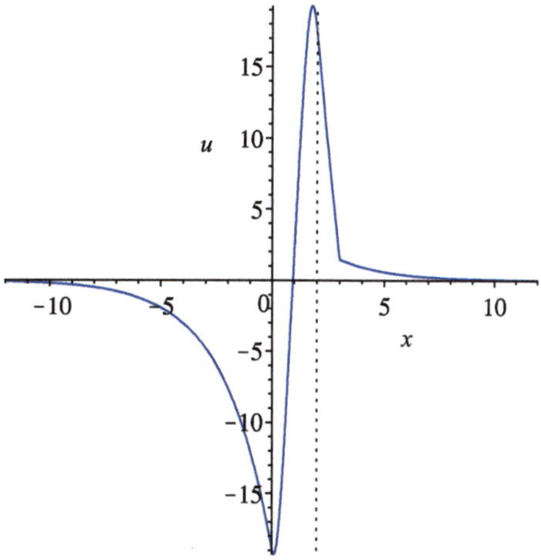

Fig. 3 The graph of u for the Kerr case. $F_R \approx 27.337$

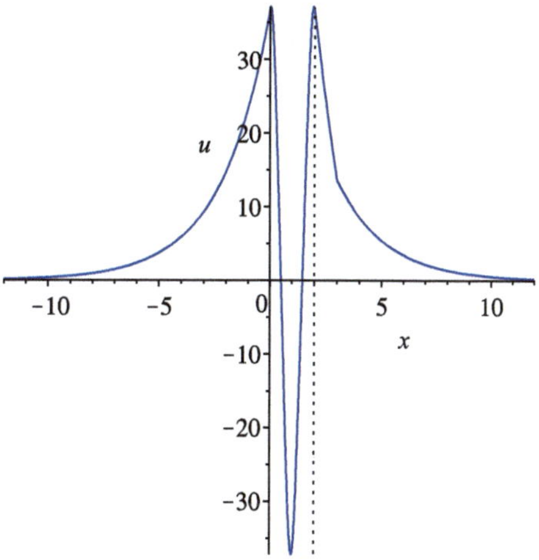

Fig. 4 The graph of u for the Kerr case. $F_R \approx 46.479$

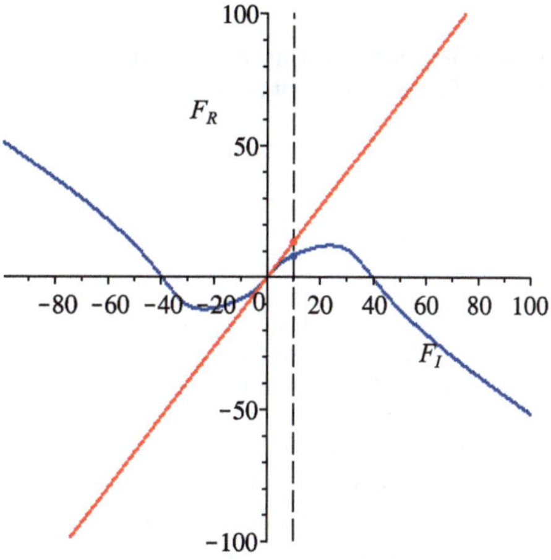

Fig. 5 The graph of curves for the "Saturated" case. Marked points of the amplitudes of the reflected field: for the linear case $F_R \approx 13.325$ (red dot); for the "Saturated" case $F_R \approx 8.187$ (blue dot)

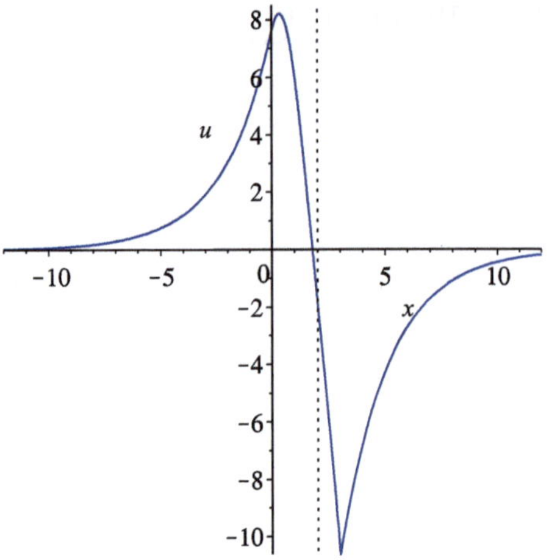

Fig. 6 The graph of u for the "Saturated" case. $F_R \approx 8.187$

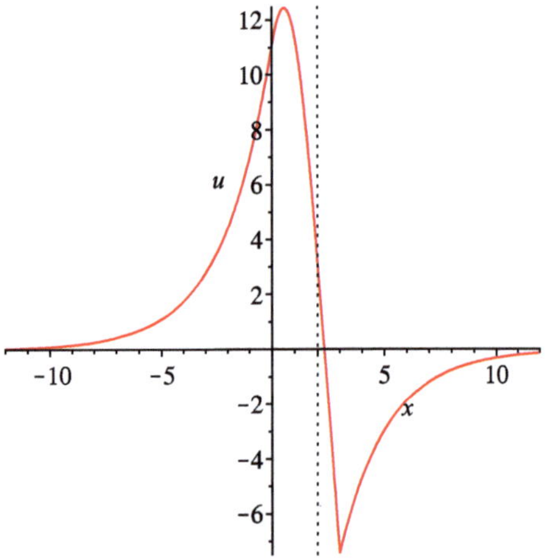

Fig. 7 The graph of u for the linear case. $F_R \approx 13.325$

Acknowledgements This study is supported by the Ministry of Education and Science of the Russian Federation [grant number 1.894.2017/4.6].

References

1. R.W.P. King, T.T. Wu, *The Scattering and Diffraction of Waves* (Harvard University Press, 1959)
2. Y. Miyazaki, Scattering and diffraction of electromagnetic waves by inhomogeneous dielectric cylinder. Radio Science **16**, 1009–1014 (1981)
3. D.V. Valovik, Integral dispersion equation method to solve a nonlinear boundary eigenvalue problem. Nonlinear Analysis: Real World Applications **20**, 52–58 (2014)
4. V. Yu. Kurseeva, D.V. Valovik, On the infinitely many electromagnetic TE eigenmodes in a plane layered waveguide filled with nonlinear medium: analytical results, in *Proceedings of the International Conference DAYS on DIFFRACTION, 2016* (2016), pp. 264–269

5. V. Yu. Kurseeva, D.V. Valovik, Theory of nonlinear guided electromagnetic waves in a plane two-layered dielectric waveguide. Math Probl Eng **2017**. Article ID 4215685, 1–19 (2017)
6. P.N. Eleonskii, L.G. Oganes'yants, V.P. Silin, Cylindrical nonlinear waveguides. Sov Phys Jetp **35**, 44–47 (1972)
7. V.A. Trenogin, *The Functional Analysis* (Nauka, Moscow, 1980)
8. I.G. Petrovsky, *Lectures on Ordinary Differential Equations Theory* (Izd. MGU, Moscow, 1984)
9. E. Smolkin, Y. Shestopalov, M. Snegur, Diffraction of TM polarized electromagnetic waves by a nonlinear inhomogeneous metal-dielectric waveguide, in *Proceedings of the 2017 19th International Conference on Electromagnetics in Advanced Applications, ICEAA 2017 8065508* (2017), pp. 1288–1291

A Nonlinear Multiparameter EV Problem

L. Angermann, Yu. V. Shestopalov, Yu. G. Smirnov and V. V. Yatsyk

Abstract We investigate a generalization of one-parameter eigenvalue problems arising in the theory of wave propagation in waveguides filled with nonlinear media to more general nonlinear multi-parameter eigenvalue problems for a nonlinear operator. Using an integral equation approach, we derive functional dispersion equations (DEs) whose roots yield the desired eigenvalues. The existence of the roots of DEs is proved and their distribution is described.

Keywords Multi-parameter eigenvalue problems · Nonlinear spectral theory · Dispersion equations

1 Introduction

In this paper we investigate nonlinear multi-parameter eigenvalue problems which constitute a new type of eigenvalue problems for nonlinear operators.

L. Angermann (✉)
Institut für Mathematik, Technische Universität Clausthal, Erzstraße 1,
38678 Clausthal-Zellerfeld, Germany
e-mail: lutz.angermann@tu-clausthal.de

Yu. V. Shestopalov
Department of Electronics, Mathematics and Natural ciences,
Faculty of Engineering and Sustainable Development,
University of Gävle, 80176 Gävle, Sweden
e-mail: yuyshv@hig.se

Yu. G. Smirnov
Department of Mathematics and Supercomputing, Penza State University,
40, Krasnaya Street, Penza 440026, Russia
e-mail: smirnovyug@mail.ru

V. V. Yatsyk
O.Ya. Usikov Institute for Radiophysics and Electronics,
National Academy of Sciences of Ukraine, 12 Ac. Proskura Street, Kharkiv 61085, Ukraine
e-mail: vasyl.yatsyk@gmail.com

L. Beilina and Yu. G. Smirnov (eds.), *Nonlinear and Inverse Problems in Electromagnetics*, Springer Proceedings in Mathematics & Statistics 243,
https://doi.org/10.1007/978-3-319-94060-1_5

Such problems arise in the theory of wave propagation in waveguides filled with nonlinear media which describes the coupled wave propagation at different frequencies [6, 7]. In these papers two-frequency wave propagation in nonlinear waveguides was considered. This paper is devoted to the n-frequency wave propagation in dielectric rod filled with nonlinear medium. This is a generalization of one-parameter eigenvalue problems considered in the theory of nonlinear waveguides [8]. The eigenvalue problems are formulated in unbounded domains, in particular, on the real axis, and with transmission-type conditions as well as conditions at infinity that contain the spectral parameters. An example of the problem statement is when the coefficient in the equation preceding the nonlinear term is nontrivial within a finite interval $(0, a)$, $a > 0$, and additional conditions are stated at the point a (continuity), at the origin (boundedness), and at infinity (rate of decay). In the papers [6, 7] the following problem has been considered: $L(\lambda_1, \lambda_2)u + \alpha B(u; \lambda_1, \lambda_2) = 0$, where L is a linear differential operator w.r.t. u and B is a nonlinear operator in all its arguments. It should be noted that the operators considered in [6, 7] do not depend explicitly on the independent variable of the problem.

In the generalization of this concept, the system of dispersion relations obtained w.r.t. several spectral parameters can be considered as a problem to determine a tuple of eigenvalues for nonlinear operator equations and constitutes in this way a nonstandard eigenvalue problem. It is worth to be mentioned that the theory of *linear* multi-parameter eigenvalue problems is well developed (see, e.g., [4]); however, there is no comparable understanding of the nonlinear situation, to the best of our knowledge.

The method of solution proposed in this study employs the transition to nonlinear integral equations [1–3, 6–8] by the help of the Green's functions of linear differential operators. After this, the eigenvalue problems are reduced to the determination of characteristic numbers of integral operator-valued functions nonlinear both with respect to the solution and spectral parameters. The latter problems are reduced to functional dispersion equations (DEs), and their roots give the desired eigenvalues. The existence of the roots is proved and their distribution is described.

The present work is aimed at developing an appropriate technique to study nonlinear multi-parameter eigenvalue problems and to prove the existence of coupled eigenvalues (eigentuples) for a generalization of the inhomogeneous nonlinear waveguide problem, with an objective to obtain the eigentuples as functions of the problem parameters (first of all, of the nonlinearity parameter). At the next step (which is not contained here and will be a subject of a separate study) we will develop and investigate a numerical method to determine the eigentuples and eigenfunctions of the problem.

2 Statement of the Problem

We consider the following system of $n \in \mathbb{N}$ differential equations

$$(\rho u_i')' + \big(\rho q_i - \lambda_i \rho - \rho^{-1} + a_i \rho f_i(|\mathbf{u}|^2)\big)u_i = 0, \quad \rho \in (0, h), \quad i = \overline{1, n}, \qquad (1)$$

where $h > 0$ is a fixed parameter, $u_i = u_i(\rho)$, and $\mathbf{u}(\rho) := \big(u_1(\rho), \ldots, u_n(\rho)\big)^\top$, $|\mathbf{u}|^2 := \sum_{j=1}^n |u_j|^2$. Here $(\,\cdot\,)^\top$ denotes the transposition operation and $i = \overline{1,n}$ means that the index i successively takes the values from 1 to n.

In this system the functions $q_i : [0, \rho] \to \mathbb{R}$ and $f_i : [0, \infty) \to \mathbb{R}$ are known continuous functions such that $q_i(\rho) > 0$ for $\rho \in [0, h]$ and

$$f_i(s) > 0 \text{ for } s > 0 \quad \text{and} \quad f_i(0) = 0 \quad \text{for every } i = \overline{1,n}.$$

The constants $a_i > 0$ are also known, whereas the vector $\lambda := (\lambda_1, \ldots, \lambda_n)^\top \in \mathbb{R}^n$ and the vector-field $\mathbf{u} : [0, h] \to \mathbb{R}^n$ are both unknown.

In addition, we assume that each function f_i is locally Lipschitzian. To be more precise, let $C[0, h]$ be the space of continuous real-valued functions on $[0, h]$ equipped with the usual norm $\|v\|_C := \max\limits_{\rho\in[0,h]} |v(\rho)|$,

$$\mathbf{C}[0, h] := \underbrace{C[0, h] \times \ldots \times C[0, h]}_{n \text{ times}}$$

be the space of continuous vector-valued functions on $[0, h]$ with the norm $\|\mathbf{v}\|_{\mathbf{C}} := \Big\{\sum\limits_{i=1}^{n} \|v_i\|_C^2\Big\}^{1/2}$, and

$$B_r := \big\{\mathbf{v} \in \mathbf{C}[0, h] : \ \|\mathbf{v}\|_{\mathbf{C}} \leqslant r\big\}$$

be a closed ball in $\mathbf{C}[0, h]$ of radius $r > 0$ with its origin at zero. Then, for each $r > 0$ there exists a constant $L_r > 0$ sucht that the following inequalities are satisfied:

$$|f_i(s_1) - f_i(s_2)| \leqslant L_r|s_1 - s_2| \quad \text{for all} \quad s_1, s_2 \in [0, r] \quad \text{and} \quad i = \overline{1,n}. \tag{2}$$

In general, the Lipschitz constant L_r depends on r.

It is convenient to arrange the n functions f_i in the form of an $n \times n$ diagonal matrix, i.e. we define $F := \operatorname{diag}(f_1, \ldots, f_n)$. By $\|F\|$ we denote the matrix norm induced by the Euclidean vector norm $|\cdot|$. In our case $\|F\| = \max_{i=\overline{1,n}} |f_i|$. Furthermore we set

$$A_r := \max_{\mathbf{v}\in B_r} \|F(|\mathbf{v}|^2)\|. \tag{3}$$

The boundary conditions for the system (1) read as follows:

$$\frac{u_i(\rho)}{\rho} \text{ is bounded on } [0, h], \quad u_i'(h) + \beta_i u_i(h) = 0, \quad i = \overline{1,n}, \tag{4}$$

where the coefficients $\beta_i = \beta_i(\lambda)$ are real-valued, continuous functions satisfying the conditions $\beta_i(\lambda) > 0$ for all $\lambda \in \boldsymbol{\Lambda}$ with a certain set $\boldsymbol{\Lambda}$ which will be specified below.

Finally we need to consider the scaling conditions

$$\mathbf{u}(h) = \mathbf{c}^+, \tag{5}$$

where $\mathbf{c}^+ := (c_1^+, \ldots, c_n^+)^\top \in \mathbb{R}^n$ is prescribed such that $c_i^+ \neq 0$ for all $i = \overline{1, n}$ (without loss of generality we may asssume that $c_i^+ > 0$ for all $i = \overline{1, n}$). We also denote $\mathbf{c}^0 := \mathbf{u}(0)$, where $\mathbf{c}^0 := (c_1^0, \ldots, c_n^0)^\top \in \mathbb{R}^n$ is unknown.

Definition 1 The vector λ is called a (scaled) eigentuple of the problem (1)–(5) if there exist nontrivial functions $u_i \in C^1(0, h] \cap C^2(0, h)$, $i = \overline{1, n}$, satisfying the system (1), the boundary conditions (4), and the scaling conditions (5). The components λ_i of λ are also called coupled eigenvalues and the functions u_i are called the eigenfunctions of the problem (1)–(5).

Remark 1 1. In contrast to the linear case (i.e. all $a_i = 0$ in (1)), where the eigenfunctions are unique up to multiplicative constants, here we have to fix the numbers c_i^+. Both the eigenvalues and the eigenfunctions depend on these numbers.
2. The property $c_i^+ = 0$ for some index i implies that $u_i = 0$. Therefore we have to assume in (5) that $c_i^+ \neq 0$ for $i = \overline{1, n}$.

Proof (of 2.) Let $c_i^+ = 0$ for some index i and suppose there is some $\varepsilon \in (0, h)$ such that $u_i(\varepsilon) \neq 0$. Then the i-th Eq. (1) reads as

$$(\rho v')' + \big(\rho q_i - \lambda_i \rho - \rho^{-1} + a_i \rho \varphi_i(v)\big)v = 0, \quad \rho \in (0, h),$$

where $v := u_i$, $\varphi_i(v) := f_i(\sum_{j=1}^{i-1} |u_j|^2 + |v|^2 + \sum_{j=i+1}^{n} |u_j|^2)$. Setting $w := v'$, this equation can be written as a first-order system

$$\begin{aligned} v' &= w, \\ w' &= -\big(\rho q_i - \lambda_i \rho - \rho^{-1} + a_i \rho \varphi_i(v)\big)v\,. \end{aligned} \tag{6}$$

The boundary conditions (4) together with the scaling conditions (5) imply that

$$v(h) = w(h) = 0. \tag{7}$$

Thus we see that the nontrivial pair $(v, w)^\top$ is a solution of the "reversed" Cauchy problem (6)–(7) on $(0, h]$ with a homogeneous "initial" condition at $\rho = h$. On the other hand, the trivial pair $(0, 0)^\top$ is also a solution of (6)–(7). Since the right-hand side of the system (6) is uniformly Lipschitz-continuous w.r.t. $(v, w)^\top$ on $[\varepsilon, h] \times [v_*, v^*] \times [w_*, w^*]$, where $v_* := \min_{\rho \in [\varepsilon, h]} v(\rho)$, $v^* := \max_{\rho \in [\varepsilon, h]} v(\rho)$ (and analogously for w_*, w^*), the so-called *main uniqueness theorem* (see, e.g., [5, Theorem 6.1]) yields a contradiction to $v(\varepsilon) \neq 0$.

The main task of this work is to prove the existence of eigentuples satisfying Definition 1.

Statement 1 *If the problem* (1)–(5) *has a solution in the sense Definition 1, then* $\lambda \in \mathbb{R}^n$ *and* $\mathbf{u}: [0,h] \to \mathbb{R}^n$.

Proof Suppose that $\mathbf{u}$ is a complex-valued vector field. Multiplying the i-th equation of (1) by $\overline{u}_i(\rho)$, then integrating and using (4), we get

$$-h\beta_i|u_i(h)|^2 - \int_0^h \rho|u_i'(\rho)|^2 d\rho + \int_0^h \rho q_i(\rho)|u_i(\rho)|^2 d\rho -$$

$$-\lambda_i \int_0^h \rho|u_i(\rho)|^2 d\rho - \int_0^h \frac{|u_i(\rho)|^2}{\rho} d\rho + a_i \int_0^h \rho f_i\big(|\mathbf{u}(\rho)|^2\big)|u_i(\rho)|^2 d\rho = 0. \quad (8)$$

Separating the imaginary part of (8) yields

$$-(\operatorname{Im}\lambda_i)\int_0^h \rho|u_i(\rho)|^2 d\rho = 0.$$

Hence $\operatorname{Im}\lambda_i = 0$ if u_i is a nontrivial function.

Furthermore, if $\lambda \in \mathbb{R}^n$, then the i-th equations in (1), (4) and (5) separate for the real and imaginary parts of u_i. Since the scaling conditions (5) are trivial for the imaginary parts, it follows by the same argument as in Remark 1(ii) that the imaginary parts are trivial.

This statement explains why it is sufficient to consider only the case of $\mathbb{R}^n$-valued λ and $\mathbf{u}$.

3 Nonlinear Integral Equations

In what follows, for the sake of brevity we will omit the phrase "for $i = \overline{1,n}$" in every formula because we will carry out our considerations for each index i.

Let us introduce n linear differential expressions

$$D_i := \frac{d}{d\rho}\Big(\rho\frac{d}{d\rho}\Big) + \Big(\rho q_i(\rho) - \lambda_i\rho - \frac{1}{\rho}\Big).$$

The expressions D_i correspond to the linear parts of Eq. (1).

We will make use of the following assumptions:

1. If λ_i is not an eigenvalue of D_i, then the Green's function G_i of the boundary value problem

$$D_i G_i = -\delta(\rho - s), \quad \partial_\rho G_i(\rho,s)|_{\rho=0} = \partial_\rho G_i(\rho,s)|_{\rho=h} = 0 \quad (9)$$

exits for such λ_i.

2. The problem
$$D_i\varphi_i = 0, \tag{10}$$
with boundary conditions
$$\varphi_i'(0) \quad \text{is bounded}, \quad \varphi_i'(h) = 0$$
has a nonempty set of simple eigenvalues
$$\ldots < \lambda_{i,j+1} < \lambda_{i,j} < \ldots < \lambda_{i,2} < \lambda_{i,1}.$$
3. In the vicinity of an eigenvalue $\lambda_i = \lambda_{i,j}$, the Green's function G_i can be written in the form
$$G_i(\rho, s; \lambda_i) = \frac{\phi_{i,j}(\rho)\phi_{i,j}(s)}{\lambda_i - \lambda_{i,j}} + \widetilde{G}_i(\rho, s; \lambda_i), \tag{11}$$
where $\phi_{i,j}(\rho)$ is an eigenfunction corresponding to an eigenvalue $\lambda_{i,j}$, $\widetilde{G}_i(\rho, s)$ is regular in the vicinity of $\lambda_{i,j}$.

Remark 2 If q_i is constant, the corresponding operator D_i satisfies these assumptions [8]. It can be proved that these assumptions are fulfilled, but we will not prove them in this paper.

Using the second Green's formula, we find that
$$\int_0^h (G_i D_i u_i - u_i D_i G_i) d\rho = \rho u_i'(\rho) G_i(\rho, s)\big|_{\rho=0}^{\rho=h} = h u_i'(h) G_i(h, s). \tag{12}$$

Using the boundary conditions (4) and the scaling conditions (5), the Eq. (12) transforms to
$$\int_0^h (G_i D_i u_i - u_i D_i G_i) d\rho = -\hat{\beta}_i c_i^+ G_i(h, s), \tag{13}$$
where $\hat{\beta}_i(\lambda) := h\beta_i(\lambda)$.

On the other hand, using (1), (9), we obtain
$$\int_0^h (G_i D_i u_i - u_i D_i G_i) d\rho = -a_i \int_0^h G_i(\rho, s)\rho f_i(|\mathbf{u}|^2) u_i(\rho) d\rho + u_i(s). \tag{14}$$

Putting (13) and (14) together, we obtain an integral representation of the solutions u_i of the system (1):

$$u_i(s) = a_i \int_0^h G_i(\rho, s)\rho f_i(|\mathbf{u}|^2)u_i(\rho)d\rho - \hat{\beta}_i c_i^+ G_i(h, s), \quad s \in [0, h]. \tag{15}$$

It is convenient to denote

$$G_i(h, 0) := \lim_{s\to 0+} G_i(h, s), \quad G_i(h, h) := \lim_{s\to h-} G_i(h, s).$$

Letting $s \to h-$ in (15), we get

$$c_i^+ = a_i \int_0^h G_i(\rho, h)\rho f_i(|\mathbf{u}|^2)u_i(\rho)d\rho - \hat{\beta}_i c_i^+ G_i(h, h). \tag{16}$$

From (16) we find

$$c_i^+ g_i(\lambda) = a_i \Phi_i(\lambda), \tag{17}$$

where

$$g_i(\lambda) := 1 + \hat{\beta}_i G(h, h),$$

$$\Phi_i(\lambda) := \int_0^h G_i(\rho, h)\rho f_i(|\mathbf{u}|^2)u_i(\rho)d\rho.$$

The solutions of the system (17) are the eigentuples of the problem (1)–(5). In order to study the solvability of the system (17), we are going to study system of nonlinear integral Eq. (15). For this reason we write the system (15) in an operator form.

We introduce the linear matrix integral operator

$$(\mathbf{Kg})(s) = \int_0^h \mathrm{K}(\rho, s)\mathbf{g}(\rho)d\rho, \quad s \in [0, h], \tag{18}$$

where $\mathbf{g} := (g_1, \dots, g_n)^\top$, the kernel $\mathrm{K}(\rho, s)$ has the form of a $n \times n$ diagonal matrix

$$\mathrm{K}(\rho, s) := \mathrm{diag}(K_{11}(\rho, s), \dots, K_{nn}(\rho, s)) \tag{19}$$

and $K_{ii}(\rho, s) := G_i(\rho, s)$.

Now the system (15) can be written in the form

$$\mathbf{u} = \mathrm{a}\mathbf{K}\big(F(|\mathbf{u}|^2)\mathbf{u}\big) + \mathbf{h}, \tag{20}$$

where $\mathrm{a} := \mathrm{diag}(a_1, \ldots, a_n)$, is an $n \times n$ diagonal matrix and the vector $\mathbf{h} := (h_1, \ldots, h_n)^\top$ has the components

$$h_i(s) := -c_i^+ \hat{\beta}_i G_i(h, s). \tag{21}$$

We shall study Eq. (20) in the space $\mathbf{C}[0, h]$.

4 Investigation of the Operator Equation

In order to prove a result about the existence of eigentuples it is necessary to formulate some auxiliary statements for Eq. (20).

The approximate solutions $\mathbf{u}^{(p)} = \big(u_1^{(p)}, \ldots, u_n^{(p)}\big)^\top$, $p = 0, 1, 2, \ldots$, of Eq. (20) can be determined using the iteration

$$\mathbf{u}^{(p+1)} = \mathrm{a}\mathbf{K}\big(F(|\mathbf{u}^{(p)}|^2)\mathbf{u}^{(p)}\big) + \mathbf{h}, \quad p = 0, 1, 2, \ldots \tag{22}$$

If the right-hand side of Eq. (20) defines a contracting operator, the sequence $(\mathbf{u}^{(p)})$ converges uniformly to a unique solution of Eq. (20). More precisely, we have the following result.

Theorem 1 *Assume there exists a number $r_0 > 0$ such that the two conditions*

$$\|\mathrm{a}\|\,\|\mathbf{K}\| A_{r_0} r_0 + \|\mathbf{h}\|_{\mathbf{C}} \leqslant r_0, \quad q := \|\mathrm{a}\|\,\|\mathbf{K}\|(2r_0^2 L_{r_0} + A_{r_0}) < 1$$

are satisfied, where the constants L_{r_0}, A_{r_0} have been defined in (2) *and* (3)*, respectively.*

Then there exists a unique solution $\mathbf{u} \in B_{r_0}$ of Eq. (20) *(or, equivalently, of the system* (15)*). The sequence of approximate solutions $(\mathbf{u}^{(p)}) \subset B_{r_0}$, which is defined by the iterative process* (22)*, converges in $\mathbf{C}[0, h]$ to a unique exact solution $\mathbf{u} \in B_{r_0}$ of Eq.* (20) *(or system* (15)*) for any initial approximation $\mathbf{u}^{(0)} \in B_{r_0}$ with the geometric rate of order q.*

Proof Consider the fixed-point equation $\mathbf{u} = \mathbf{A}(\mathbf{u})$ in the space $\mathbf{C}[0, h]$ with the nonlinear operator

$$\mathbf{A}(\mathbf{u}) := \mathrm{a}\mathbf{K}\big(F(|\mathbf{u}|^2)\mathbf{u}\big) + \mathbf{h}\,.$$

Since

$$\|\mathbf{A}(\mathbf{u})\|_{\mathrm{C}} = \big\|\mathrm{a}\mathbf{K}\big(F(|\mathbf{u}|^2)\mathbf{u}\big) + \mathbf{h}\big\|_{\mathrm{C}} \leqslant \|\mathrm{a}\|\,\|\mathbf{K}\|A_{r_0}r_0 + \|\mathbf{h}\|_{\mathrm{C}} \quad \text{for all } \mathbf{u} \in B_{r_0}, \tag{23}$$

the operator $\mathbf{A}$ maps the ball B_{r_0} into itself.

Next we show that the following estimate is valid for arbitrary $\mathbf{u}, \mathbf{v} \in B_{r_0}$:

$$\begin{aligned}\|\mathbf{A}(\mathbf{u}) - \mathbf{A}(\mathbf{v})\|_{\mathrm{C}} &= \big\|\mathrm{a}\mathbf{K}\big(F(|\mathbf{u}|^2)\mathbf{u} - F(|\mathbf{v}|^2)\mathbf{v}\big)\big\|_{\mathrm{C}} \leqslant \\ &\leqslant \|\mathrm{a}\|\,\|\mathbf{K}\|(2r_0^2L_{r_0} + A_{r_0})\|\mathbf{u} - \mathbf{v}\|_{\mathrm{C}}. \end{aligned} \tag{24}$$

Indeed,

$$\begin{aligned}&\big\|F(|\mathbf{u}|^2)\mathbf{u} - F(|\mathbf{v}|^2)\mathbf{v}\big\|_{\mathrm{C}} = \\ &\quad= \big\|\big(F(|\mathbf{u}|^2)\mathbf{u} - F(|\mathbf{v}|^2)\mathbf{u}\big) + \big(F(|\mathbf{v}|^2)\mathbf{u} - F(|\mathbf{v}|^2)\mathbf{v}\big)\big\|_{\mathrm{C}} \leqslant \\ &\quad\leqslant \big\|F(|\mathbf{u}|^2)\mathbf{u} - F(|\mathbf{v}|^2)\mathbf{u}\big\|_{\mathrm{C}} + \big\|F(|\mathbf{v}|^2)\mathbf{u} - F(|\mathbf{v}|^2)\mathbf{v}\big\|_{\mathrm{C}} \leqslant \\ &\quad\leqslant \big\|F(|\mathbf{u}|^2) - F(|\mathbf{v}|^2)\big\|\,\|\mathbf{u}\|_{\mathrm{C}} + \big\|F(|\mathbf{v}|^2)\big\|\,\|\mathbf{u} - \mathbf{v}\|_{\mathrm{C}} \leqslant \\ &\quad\leqslant L_{r_0}\big\||\mathbf{u}|^2 - |\mathbf{v}|^2\big\|\,\|\mathbf{u}\|_{\mathrm{C}} + \big\|F(|\mathbf{v}|^2)\big\|\,\|\mathbf{u} - \mathbf{v}\|_{\mathrm{C}} \leqslant \\ &\quad\leqslant L_{r_0}\big\||\mathbf{u}| - |\mathbf{v}|\big\|_C\big\||\mathbf{u}| + |\mathbf{v}|\big\|_C\|\mathbf{u}\|_{\mathrm{C}} + \big\|F(|\mathbf{v}|^2)\big\|\,\|\mathbf{u} - \mathbf{v}\|_{\mathrm{C}}.\end{aligned}$$

By the left-hand side of the triangle inequality, that is $\big\||\mathbf{u}| - |\mathbf{v}|\big\|_C \leqslant \|\mathbf{u} - \mathbf{v}\|_{\mathrm{C}}$, we get $\big\|F(|\mathbf{u}|^2)\mathbf{u} - F(|\mathbf{v}|^2)\mathbf{v}\big\|_{\mathrm{C}} \leqslant (2r_0^2L_{r_0} + A_{r_0})\|\mathbf{u} - \mathbf{v}\|_{\mathrm{C}}$. This implies (24).

In summary we have shown that the operator $\mathbf{A}$ is contracting in the closed ball B_{r_0} and the statement follows from Banach's fixed-point theorem.

For further study we consider the auxiliary equation

$$\|\mathrm{a}\|\,\|\mathbf{K}\|\,\|F(r^2)\|r + \|\mathbf{h}\| = r. \tag{25}$$

To do so, we introduce the function $y(r) := r - \|\mathrm{a}\|\,\|\mathbf{K}\|\,\|F(r^2)\|r$, $r \in [0, \infty)$, and consider the following two cases:

(a) y is a monotonically increasing function for $r \in [0, \infty)$;
(b) All entries f_i of F are differentiable and y is a nonmonotonic function for $r \in [0, \infty)$.

In the case **(a)**, Eq. (25) has a unique positive solution $r = r_*$ for any positively definite matrix a and any vector $\mathbf{h}$ satisfying the inequality $0 < \|\mathbf{h}\|_{\mathrm{C}} < y_{\max}$, where $y_{\max} := \lim_{r\to\infty} y(r)$ provided the limit exists and $y_{\max} := \infty$ otherwise. Obviously, the value $y_{\max}$ depends on the matrix a.

In the case **(b)**, we consider the equation $y'(r) = 0$. Due to the nonmonotonic behaviour of y and to the Lipschitz-continuity of f_i there exists a minimal positive root r_m of this equation. Since $y'(0) = 1$, the function y monotonically increases for $r \in [0, r_m]$ having a maximum $y^*_{\max}$ at the point $r = r_m$. This means that Eq. (25) has a unique positive solution $r = r_*$ for any positive matrix a and any vector $\mathbf{h}$ satisfying the inequality $0 < \|\mathbf{h}\|_C < y^*_{\max}$. Obviously, the value $y^*_{\max}$ depends on matrix a.

Statement 2 **(case (a))** *Let r_* be an arbitrary positive constant. If the matrix* a *is such that* $\|\text{a}\| < A$, *where* $A := \left(\|\mathbf{K}\|(2r_*^2 L_{r_*} + A_{r_*})\right)^{-1}$, *and the vector* $\mathbf{h}$ *satisfies the inequality* $0 < \|\mathbf{h}\|_C < y_{\max}$, *then the Eq.* (20) *has a unique solution* $\mathbf{u} \in B_{r_*}$.

Proof Let $\mathbf{u} \in B_{r_*}$, then the following estimate is true

$$\|\mathbf{A}(\mathbf{u})\|_C = \left\|\text{a}\mathbf{K}\big(f(|\mathbf{u}|^2)\mathbf{u}\big) + \mathbf{h}\right\|_C \leqslant \|\text{a}\|\,\|\mathbf{K}\|A_{r_*}r_* + \|\mathbf{h}\|_C = r_*. \tag{26}$$

Hence if $\mathbf{u}, \mathbf{v} \in B_{r_*}$, then the estimate (24) is valid with $r_0 := r_*$.

Since $\|\text{a}\| < A$, the inequality $\|\text{a}\|\,\|\mathbf{K}\|(2r_*^2 L_{r_*} + A_{r_*}) < 1$ is valid. Thus the conditions of Theorem 1 are satisfied with $q := \|\text{a}\|\,\|\mathbf{K}\|(2r_*^2 L_{r_*} + A_{r_*})$.

We conclude that $\mathbf{A}$ maps B_{r_*} into itself and is a contracting operator in B_{r_*}. This implies the existence of a unique solution to Eq. (20) in the ball B_{r_*}.

Statement 3 **(case (b))** *Let $r_* := r_m$. If the matrix* a *is such that* $\|\text{a}\| < A$, *where* $A := \left(\|\mathbf{K}\|(2r_*^2 L_{r_*} + A_{r_*})\right)^{-1}$, *and the vector* $\mathbf{h}$ *satisfies the inequality* $0 < \|\mathbf{h}\|_C < y^*_{\max}$, *then the Eq.* (20) *has a unique solution* $\mathbf{u} \in B_{r_*}$.

Statements 3 is proved in the same way as Statement 2.

Remark 3 In the cases **(a)**, **(b)**, the constants L_{r_*} and A_{r_*} depend on r_*.

The following step is to prove the continuous dependence of the solution $\mathbf{u} = \mathbf{u}(\rho; \lambda)$ on the parameter λ if the parameter belongs to a certain set $\boldsymbol{\Lambda} \subset \mathbb{R}^n$.

Theorem 2 *Let the matrix integral operator* $\mathbf{K} = \mathbf{K}(\lambda)$ *and the vector* $\mathbf{h} = \mathbf{h}(\lambda)$ *of Eq.* (20) *depend continuously on the parameters* $\lambda \in \boldsymbol{\Lambda}$ *for a certain closed real n-dimensional set* $\boldsymbol{\Lambda} := \Lambda_1 \times \ldots \times \Lambda_n \subset \mathbb{R}^n$. *Let also one of the Statements 2 or 3 be valid. Then the solution* $\mathbf{u} = \mathbf{u}(\rho; \lambda)$ *of Eq.* (20) *exists for* $\rho \in [0, h]$, *is unique and continuously depends on* $\lambda \in \boldsymbol{\Lambda}$.

Proof It follows from Theorem 1 that there exists a unique nontrivial solution $\mathbf{u} = \mathbf{u}(\rho, \lambda)$ of Eq. (20). Now we prove that this solution depends continuously on the parameter $\lambda \in \boldsymbol{\Lambda}$.

Choosing r_*, we define $L_{r_*} := L_{r_*}(\lambda)$ and $A_{r_*} := A_{r_*}(\lambda)$, which continuously depend on λ. We stress that in general the value r_* depends on λ in the case **(b)**. However, fixing the set $\boldsymbol{\Lambda}$, we can choose the maximal possible value r_* (and thus r_* no longer depends on $\boldsymbol{\Lambda}$). Denote $L_{\max} := \max\limits_{\lambda \in \boldsymbol{\Lambda}} L_{r_*}(\lambda)$ and $A_{\max} := \max\limits_{\lambda \in \boldsymbol{\Lambda}} A_{r_*}(\lambda)$.

As in the proof of Theorem 1 it follows from the above inequality that

$$\left\| F(|\mathbf{u}|^2)\mathbf{u} - F(|\mathbf{v}|^2)\mathbf{v} \right\|_{\mathbf{C}} \leqslant (2r_*^2 L_{\max} + A_{\max}) \|\mathbf{u} - \mathbf{v}\|_{\mathbf{C}}.$$

First, let us assume that

$$\|\mathbf{u}(\lambda)\|_{\mathbf{C}} \geqslant \|\mathbf{u}(\lambda + \Delta\lambda)\|_{\mathbf{C}}\,. \tag{27}$$

Then the following inequalities are valid

$$\begin{aligned}
&|\mathbf{u}(s;\lambda+\Delta\lambda) - \mathbf{u}(s;\lambda)| = \\
&\quad = \left| \int_0^h \mathrm{K}(\rho,s;\lambda+\Delta\lambda) F\big(|\mathbf{u}(\rho;\lambda+\Delta\lambda)|^2\big)\mathbf{u}(\rho;\lambda+\Delta\lambda)dx \right. - \\
&\quad \left. - \int_0^h \mathrm{K}(\rho,s;\lambda) F\big(|\mathbf{u}(\rho;\lambda)|^2\big)\mathbf{u}(\rho;\lambda)dx + \mathbf{h}(s;\lambda+\Delta\lambda) - \mathbf{h}(s;\lambda) \right| \leqslant \\
&\quad \leqslant \left| \int_0^h \big\{\mathrm{K}(\rho,s;\lambda+\Delta\lambda) - \mathrm{K}(\rho,s;\lambda)\big\} F\big(|\mathbf{u}(\rho;\lambda+\Delta\lambda)|^2\big)\mathbf{u}(\rho;\lambda+\Delta\lambda)dx \right| + \\
&+ \left| \int_0^h \mathrm{K}(\rho,s;\lambda) \Big\{ F\big(|\mathbf{u}(\rho;\lambda+\Delta\lambda)|^2\big)\mathbf{u}(\rho;\lambda+\Delta\lambda) - F\big(|\mathbf{u}(\rho;\lambda)|^2\big)\mathbf{u}(\rho;\lambda) \Big\} dx \right| + \\
&\qquad + |\mathbf{h}(s;\lambda+\Delta\lambda) - \mathbf{h}(s;\lambda)|.
\end{aligned}$$

It follows from the previous lines that

$$\begin{aligned}
&\|\mathbf{u}(\lambda+\Delta\lambda) - \mathbf{u}(\lambda)\|_{\mathbf{C}} \leqslant \|\mathrm{a}\| \, \|\mathbf{K}(\lambda+\Delta\lambda) - \mathbf{K}(\lambda)\| A_{\max} r_* + \\
&+ \|\mathrm{a}\| \, \|\mathbf{K}(\lambda)\| (2r_*^2 L_{\max} + A_{\max}) \|\mathbf{u}(\lambda+\Delta\lambda) - \mathbf{u}(\lambda)\|_{\mathbf{C}} + \|\mathbf{h}(\lambda+\Delta\lambda) - \mathbf{h}(\lambda)\|_{\mathbf{C}}.
\end{aligned}$$

Thus, we obtain that

$$\begin{aligned}
&\|\mathbf{u}(\lambda+\Delta\lambda) - \mathbf{u}(\lambda)\|_{\mathbf{C}} \leqslant \\
&\qquad \leqslant \frac{\|\mathrm{a}\| \, \|\mathbf{K}(\lambda+\Delta\lambda) - \mathbf{K}(\lambda)\| A_{\max} r_* + \|\mathbf{h}(\lambda+\Delta\lambda) - \mathbf{h}(\lambda)\|_{\mathbf{C}}}{1 - \|\mathrm{a}\| (2r_*^2 L_{\max} + A_{\max}) \max\limits_{\lambda \in \Lambda} \|\mathbf{K}(\lambda)\|},
\end{aligned} \tag{28}$$

where the denominator in the right-hand side is strongly positive and does not depend on λ. Obviously, the numerator in the right-hand side tends to zero as $\Delta\lambda \to 0$.

Now, let $\|\mathbf{u}(\lambda)\|_{\mathbf{C}} \leqslant \|\mathbf{u}(\lambda + \Delta\lambda)\|_{\mathbf{C}}$. Then, all the preceding estimations remain valid if we replace λ by $\lambda + \Delta\lambda$ and $\lambda + \Delta\lambda$ by λ. Thus the estimate (28) also remains valid.

5 Existence of Eigentuples

In order to prove the solvability of DEs (17), at first we study the corresponding linear problem.

5.1 *The Linear Problem*

Setting a $= 0$, we get from (17) that

$$g_i(\lambda) = 0 \quad \text{for all } i = \overline{1, n}. \tag{29}$$

Using (11), we obtain

$$G_i(h, h; \lambda_i) = \frac{\phi_{i,j}^2(h)}{\lambda_i - \lambda_{i,j}} + \widetilde{G}_i(h, h; \lambda_i).$$

It is easy to show that $\phi_{i,j}(h) \neq 0$ for $i = \overline{1, n}$ and for all j. Indeed, if $\phi_{i,j}(h) = 0$, then it follows from (4) that $\phi_{i,j}'(h) = 0$. So we get a Cauchy problem for the equation $D_i u = 0$ with zero initial data, and it follows by the same argument as in Remark 1(ii) that this equation has only the trivial solution. But this contradicts to the statement that an eigenfunction $\phi_{i,j}$ is not equal identically to zero on $[0, h]$.

Multiplying $g_i(\lambda)$ by $\lambda_i - \lambda_{i,l}$, we get

$$(\lambda_i - \lambda_{i,j})g_i(\lambda) = (\lambda_i - \lambda_{i,j})\big(1 + \hat{\beta}_i\widetilde{G}_i(h, h; \lambda_i)\big) + \hat{\beta}_i\phi_{i,j}^2(h).$$

Since $\beta_i^2(\lambda) > 0$ and $\phi_{i,l}^2(h) > 0$, we have

$$\lim_{\lambda_i \to \lambda_{i,j}} \big((\lambda_i - \lambda_{i,j})g_i(\lambda)\big) = \hat{\beta}_i\phi_{i,j}^2(h) > 0.$$

Then,

$$\lim_{\lambda_i \to \lambda_{i,j} \pm 0} g_i(\lambda) = \pm\infty$$

and, taking into account that the function $g_i(\lambda)$ is continuous in the interval $(\lambda_{i,l+1}, \lambda_{i,l})$ for all $l \geqslant 1$, we obtain that, between $\lambda_{i,l+1}$ and $\lambda_{i,l}$, there is at least one root $\widetilde{\lambda}_{i,l}$ of the equation $g_i(\lambda) = 0$. Obviously, this is true for each $i = \overline{1, n}$.

In other words, for each $i = \overline{1, n}$ and any integer $j \geqslant 1$, there exist $\widetilde{\lambda}_{i,j}$ such that the function $g_i(\lambda)$ can be represented in the form

$$g_i(\lambda) = (\lambda_i - \widetilde{\lambda}_{i,j})\widetilde{g}_i(\lambda),$$

where $\widetilde{\lambda}_{i,j}$ depends on $\lambda_1, \ldots, \lambda_{i-1}, \lambda_{i+1}, \ldots, \lambda_n$. Thus we have proved the following result.

Theorem 3 *If the problem* (10) *has* $p_i + 1$ *eigenvalues, then the problem* (1)–(4) *with* $\mathrm{a} = 0$ *has at least* $p_1 \times \ldots \times p_n$ *isolated eigentuples*

$$\widetilde{\lambda}_{j_1,\ldots,j_n} = (\widetilde{\lambda}_{1,j_1}, \ldots, \widetilde{\lambda}_{n,j_n})^\top,$$

where $j_k = 1, 2, \ldots, k = \overline{1, n}$.

All the eigentuples are real-valued vectors and, for each i, *the* i*-th component of an eigentuple is a simple root of* $g_i(\lambda) = 0$.

5.2 The Nonlinear Problem

The following theorem is the main result of this study.

Theorem 4 *Let* $\boldsymbol{\Lambda} = \Lambda_1 \times \cdots \times \Lambda_n$ *be a bounded set, where each of* Λ_i *is a segment on the real axis, such that Theorem 2 takes place, one of the Statements 2 or 3 holds, and the linear problem has* $p_1 \times \ldots \times p_n$ *eigentuples*

$$\widetilde{\lambda}_{j_1,\ldots,j_n} = (\widetilde{\lambda}_{1,j_1}, \ldots, \widetilde{\lambda}_{n,j_n})$$

inside $\boldsymbol{\Lambda}$ *and each* $\widetilde{\lambda}_{i,j_k} \in \Lambda_i$, *where* $j_1 = \overline{1, p_1}, \ldots, j_n = \overline{1, p_n}$.

Then there exists a matrix $\mathrm{a}^0 = \operatorname{diag}(a_1^0, \ldots, a_n^0)$, *where each* $a_i^0 > 0$, *such that for any matrix* $\mathrm{a} = \operatorname{diag}(a_1, \ldots, a_n)$, *where* $a_i \leqslant a_i^0$, $i = \overline{1, n}$, *the problem* (1)–(5) *has at least* $p_1 \times \ldots \times p_n$ *eigentuples*

$$\widehat{\lambda}_{j_1,\ldots,j_n} = (\widehat{\lambda}_{1,j_1}, \ldots, \widehat{\lambda}_{n,j_n})$$

such that each $\widehat{\lambda}_{i,j_k} \in \Lambda_i$.

Proof Let $\widetilde{\lambda}_{j_1,\ldots,j_n} := (\widetilde{\lambda}_{1,j_1}, \ldots, \widetilde{\lambda}_{n,j_n})$ be a solution to the linear problem. Then $\widetilde{\lambda}_{i,j_k} \in \Lambda_i$ for $i = \overline{1, n}$. Let the positive values $\delta^{\pm}_{i,j_k}$ be sufficiently small such that

$$[\widetilde{\lambda}_{i,j_k} - \delta^{-}_{i,j_k}, \widetilde{\lambda}_{i,j_k} + \delta^{+}_{i,j_k}] =: \Delta_{i,j_k} \subset \Lambda_i \quad \text{for } i = \overline{1, n}.$$

The quantities $\delta^{\pm}_{i,j_k}$ are chosen in the way that for all possible i and j_k the segments Δ_{i,j_k} do not contain poles of the Green's function $G_i(\rho, s; \lambda_i)$, see (9), (11).

Introduce the notation

$$\Delta_{k_1,\ldots,k_n} := \Delta_{1,k_1} \cup \ldots \cup \Delta_{n,k_n},$$

where $(k_1, \ldots, k_n)$ is a particular n-tuple such that $k_i \in \{1, \ldots, p_i\}$.

Let us consider $A := \left(\|\mathbf{K}\|(2r_*^2 L_{r_*} + A_{r_*})\right)^{-1}$. It is clear that $A = A(\lambda)$. Introduce the numbers

$$A_{k_1,\ldots,k_n} := \min_{\lambda \in \Delta_{k_1,\ldots,k_n}} A(\lambda).$$

Then choosing a positively definite matrix a^0 such that $\|\mathrm{a}^0\| < A_{k_1,\dots,k_n}$, in accordance with one of the Statements 2 or 3, a unique solution $\mathbf{u} = \mathbf{u}(s, \lambda)$ to Eq. (20), defined globally on $s \in [0, h]$, exists for all $\lambda \in \Delta_{k_1,\dots,k_n}$. This solution is continuous in s, λ and $\|\mathbf{u}\|_C \leqslant r_* = r_*(\lambda)$, where r_* is defined in one of the Statements 2 or 3. Let $r_{k_1,\dots,k_n} := \max_{\lambda \in \Delta_{k_1,\dots,k_n}} r_*$. Estimating Φ_i, we see that

$$\max_{\lambda \in \Delta_{k_1,\dots,k_n}} |\Phi_i(\lambda)| \leqslant c_{k_1,\dots,k_n} r_{k_1,\dots,k_n} \max_{\lambda \in \Delta_{k_1,\dots,k_n}} A_{r_*},$$

where $c_{k_1,\dots,k_n}$ is a positive constant independent of λ or any of its components.

It follows from the proof of Theorem 3 that each of the functions $g_i(\lambda)$ is continuous and has different signs at $\lambda_i = \widetilde{\lambda}_{i,j_k} - \delta^-_{i,j_k}$ and $\lambda_i = \widetilde{\lambda}_{i,j_k} + \delta^+_{i,j_k}$ and

$$(\lambda_1, \dots, \lambda_{i-1}, \lambda_{i+1}, \dots, \lambda_n) \subset \Delta_{1,k_1} \cup \dots \cup \Delta_{i-1,k_{i-1}} \cup \Delta_{i+1,k_{i+1}} \cup \dots \cup \Delta_{n,k_n}$$

for all possible i and $(n-1)$-tuples $(k_1, \dots, k_{i-1}, k_{i+1}, \dots, k_n)$.

Introduce the numbers

$$m_{k_1,\dots,k_n} := \max_{\lambda \in \Delta_{k_1,\dots,k_n}} |g_i(\lambda)|.$$

Now it is clear that if a^0 is such that

$$\|\mathrm{a}^0\| \leqslant \frac{m_{k_1,\dots,k_n}}{c_{k_1,\dots,k_n} r_{k_1,\dots,k_n} \max_{\lambda \in \Delta_{k_1,\dots,k_n}} A_{r_*}},$$

then

$$\big(c_i^+ g_i(\lambda^{(1)}) - a_i \Phi(\lambda^{(1)})\big)\big(c_i^+ g_i(\lambda^{(2)}) - a_i \Phi(\lambda^{(2)})\big) < 0 \tag{30}$$

for all $a_i \leqslant a_i^0$ and $i = \overline{0, n}$, where $\lambda^{(1)} := (\lambda_1^{(1)}, \dots, \lambda_n^{(1)})$, $\lambda^{(2)} := (\lambda_1^{(2)}, \dots, \lambda_n^{(2)})$; $\lambda_i^{(1)}$ is equal either $\widetilde{\lambda}_{i,j_k} - \delta^-_{i,j_k}$ or $\widetilde{\lambda}_{i,j_k} + \delta^+_{i,j_k}$ and $\lambda_i^{(2)}$ is equal either $\widetilde{\lambda}_{i,j_k} + \delta^+_{i,j_k}$ or $\widetilde{\lambda}_{i,j_k} - \delta^-_{i,j_k}$, respectively. Due to continuous dependence of the left-hand side of (30) on each parameter, the latter inequality implies the existence of n-tuple $\widehat{\lambda} = (\widehat{\lambda}_1, \dots, \widehat{\lambda}_n)$ in each $\Delta_{k_1,\dots,k_n}$ such that

$$c_i^+ g_i(\widehat{\lambda}) - a_i \Phi(\widehat{\lambda}) = 0.$$

Thus the found $\widehat{\lambda}$ is a solution to (1)–(5), that is, an eigentuple of the nonlinear problem. Then we can choose a^0 in such a way that

$$\|\mathrm{a}^0\| \leqslant \min\left\{A_{k_1,\dots,k_n}, \frac{m_{k_1,\dots,k_n}}{c_{k_1,\dots,k_n} r_{k_1,\dots,k_n} \max_{\lambda \in \Delta_{k_1,\dots,k_n}} A_{r_*}}\right\}.$$

Obviously, the procedure described above is valid for each $\Delta_{k_1,\ldots,k_n}$. This implies the existence of at least one solution to the nonlinear problem in the vicinity of each solution to the linear problem in the fixed domain $\boldsymbol{\Lambda}$. Thus there exist at least $p_1 \times \ldots \times p_n$ eigentuples $\widehat{\lambda}_{j_1,\ldots,j_n} = (\widehat{\lambda}_{1,j_1}, \ldots, \widehat{\lambda}_{n,j_n})$, where $j_1 = \overline{1, p_1}, \ldots, j_n = \overline{1, p_n}$.

Carrying out the above estimates for all n-tuples $k_1, \ldots, k_n$, we obtain the estimate for a^0, which is valid for $\boldsymbol{\Delta} := \bigcup_{k_1,\ldots,k_n} \Delta_{k_1,\ldots,k_n}$.

6 Conclusion

We have constructed a generalization of one-parameter eigenvalue problems that arise particularly in the theory of wave propagation in waveguides filled with nonlinear media to more general statements in the form of nonlinear multi-parameter eigenvalue problems. We have introduced the notion of an eigentuple strengthening that, within the frames of multi-parameter eigenvalue problem statement, one looks for finite sets of eigenvalues rather than for separate numbers. The multi-parameter setting has been reformulated for a well-defined family of nonlinear operators admitting the application of the contraction principle. Using a reduction to integral equations, we have obtained functional DEs and demonstrated that the roots of these DEs are the desired eigenvalues. The developed approach, which both offers a specific tool to calculate the eigenvalues and a method to prove their existence and describe distribution, is a separate important result of this study.

Acknowledgments We acknowledge the support by the Ministry of Education and Science of the Russian Federation (agreement no. 1.894.2017), by Mathematisches Forschungsinstitut Oberwolfach ("Research in Pairs" stay 2014).

References

1. L. Angermann, V.V. Yatsyk, Resonance properties of scattering and generation of waves on cubically polarisable dielectric layers, in *Electromagnetic Waves*, ed. by V. Zhurbenko (InTech, Rijeka/Vienna, Croatia/Austria, 2011), pp. 299–340
2. L. Angermann, Y.V. Shestopalov, V.V. Yatsyk, Mathematical models for scattering and generation of plane wave packets on layered, cubically polarisable structures. Far East J. Appl. Math. **81**(1–2), 1–31 (2013)
3. L. Angermann, V.V. Yatsyk, Mathematical models of electrodynamical processes of wave scattering and generation on cubically polarisable layers. Prog. Electromagn. Res. B **56**, 109–136 (2013)
4. E.V. Atkinson, A.B. Mingarelli, *Multiparameter Eigenvalue Problems: Sturm-Liouville Theory* (CRC Press, Boca Raton, 2011)
5. G. Birkhoff, G.-C. Rota, *Ordinary Differential Equations*, 4th edn. (Wiley, New York-Chichester, 1989)

6. Y.G. Smirnov, D.V. Valovik, Coupled electromagnetic transverse-electric-transverse magnetic wave propagation in a cylindrical waveguide with Kerr nonlinearity. J. Math. Phys. **54**(4), 043506–1–22 (2013)
7. Y.G. Smirnov and D.V. Valovik. Problem of nonlinear coupled electromagnetic TE-TE wave propagation. J. Math. Physics, **54**(8), 083502–1–13 (2013)
8. Y.G. Smirnov, H.W. Schürmann, Y.V. Shestopalov, Integral equation approach for the propagation of TE-waves in a nonlinear dielectric cylinrical waveguide. J. Nonlinear Math. Phys. **11**(2), 256–268 (2004)

Numerical Study of the Azimuthal Symmetric Hybrid Waves in a Nonlinear Cylindrical Waveguide

E. Smolkin

Abstract The propagation of monochromatic nonlinear symmetric hybrid waves in a cylindrical nonlinear dielectric waveguide is considered. The physical problem is reduced to solving a transmission eigenvalue problem for a system of ordinary differential equations. Spectral parameters of the problem are propagation constants of the waveguide. Numerical results are found with the modification of the shooting method. The method allows us to determine approximate eigenvalues with any prescribed accuracy. The comparison with known exact solutions (for particular values of parameters) are made. The approach described in this paper can be applied to other problems, e.g., to multilayered inhomogeneous waveguides.

Keywords Waveguide · Kerr law · Nonlinear hybrid wave · Numerical method

1 Introduction

Linear theory of the wave propagation in dielectric waveguides of simple geometry (circular cylindrical and planar layered) describing polarized (TE and TM) electromagnetic waves has been known for years. Theory of propagation of electromagnetic waves in such waveguides is well elaborated in the linear case, i.e. when the waveguide is filled with linear isotropic homogeneous medium [1, 2].

It is known that the permittivity of a dielectric may depend nonlinearly on the electromagnetic field intensity. For this reason nonlinear (with Kerr nonlinearity) waveguide structures were studied [3], where the surface waves propagating along the waveguide depend harmonically on the longitudinal coordinate z and do not depend on azimuthal angle φ. Short background of nonlinear guided waves in media with Kerr and Kerr-like nonlinearities is given in Ref. [4]. For purely nonlinear TE and TM waves see Refs. [5–8].

E. Smolkin (✉)
Penza State University, 40 Krasnaya Street, Penza 440026, Russia
e-mail: e.g.smolkin@hotmail.com

L. Beilina and Yu. G. Smirnov (eds.), *Nonlinear and Inverse Problems in Electromagnetics*, Springer Proceedings in Mathematics & Statistics 243,
https://doi.org/10.1007/978-3-319-94060-1_6

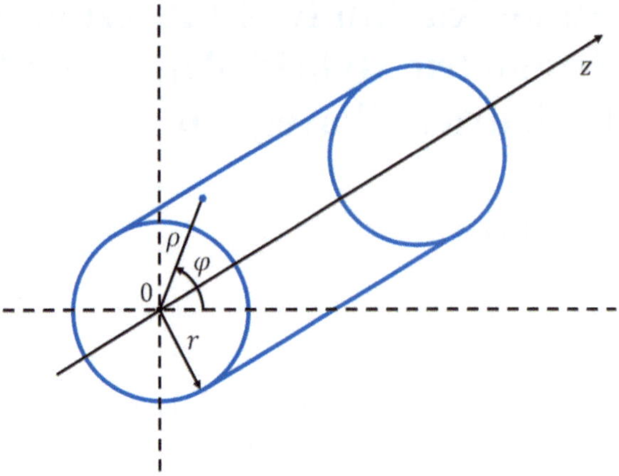

Fig. 1 The cylindrical waveguide Σ

In this work we consider a circular dielectric waveguide filled with nonlinear isotropic inhomogeneous medium (Fig. 1). We will assume that medium in the waveguide is described by the Kerr law, i.e. permittivity depends on the electric field as follows: $\varepsilon = \varepsilon_0\varepsilon_2(\rho) + \varepsilon_0\alpha\,|\mathbf{E}|^2$.

A numerical method is proposed for determination of the propagation constants for TE and TM cases can be applied for numerical study of propagation of hybrid waves [9, 10]. Nonlinear symmetric hybrid waves are very interesting because they have no analogues among the linear waves. It is a new class of nonlinear waves.

A similar study was carried out for a homogeneous dielectric layer, inhomogeneous metal-dielectric waveguides [11–13].

2 Statement of the Problem

Consider three-dimensional space $\mathbb{R}^3$ with a cylindrical coordinate system $O\rho\varphi z$ filled with isotropic medium having constant permittivity $\varepsilon_1\varepsilon_0$ ($\varepsilon_0 > 0$ is the permittivity of free space), and constant permeability $\mu = \mu_0$ (where $\mu_0 > 0$ is the permeability of free space).

A cylindrical waveguide Σ filled with nonlinear medium is placed parallel to the axis Oz. The waveguide Σ has a cross section

$$\Sigma := \{(\rho, \varphi, z) : 0 \le \rho \le r, 0 \le \varphi < 2\pi\}$$

and its generating line (the waveguide axis) is parallel to the axis Oz (see. Fig. 1).

Maxwell's equations have the form [14]

$$\begin{aligned} \mathrm{rot}\tilde{\mathbf{H}} &= \partial_t\tilde{\mathbf{D}}, \\ \mathrm{rot}\tilde{\mathbf{E}} &= -\partial_t\tilde{\mathbf{B}}, \end{aligned} \tag{1}$$

where $\widetilde{\mathbf{D}} = \varepsilon\widetilde{\mathbf{E}}$, $\widetilde{\mathbf{B}} = \mu_0\widetilde{\mathbf{H}}$ and $\partial_t \equiv \partial/\partial t$. From system (1) we obtain

$$\begin{aligned} \mathrm{rot}\tilde{\mathbf{E}} &= -\partial_t(\mu_0\tilde{\mathbf{H}}), \\ \mathrm{rot}\tilde{\mathbf{H}} &= \partial_t(\varepsilon\tilde{\mathbf{E}}). \end{aligned} \tag{2}$$

We assume that the fields $\tilde{\mathbf{E}}$, $\tilde{\mathbf{H}}$ depend harmonically on time [3]

$$\begin{aligned} \widetilde{\mathbf{E}}(\rho,\varphi,z,t) &= \mathbf{E}^+(\rho,\varphi,z)\cos\omega t + \mathbf{E}^-(\rho,\varphi,z)\sin\omega t, \\ \widetilde{\mathbf{H}}(\rho,\varphi,z,t) &= \mathbf{H}^+(\rho,\varphi,z)\cos\omega t + \mathbf{H}^-(\rho,\varphi,z)\sin\omega t, \end{aligned}$$

where ω is the circular frequency and $\widetilde{\mathbf{E}}, \mathbf{E}^+, \mathbf{E}^-, \widetilde{\mathbf{H}}, \mathbf{H}^+, \mathbf{H}^-$ are the real functions.

It is clear that

$$\widetilde{\mathbf{E}} = \mathrm{Re}\left\{\mathbf{E}e^{-i\omega t}\right\}, \quad \widetilde{\mathbf{H}} = \mathrm{Re}\left\{\mathbf{H}e^{-i\omega t}\right\},$$

where

$$\mathbf{E} = \mathbf{E}^+ + i\mathbf{E}^-, \quad \mathbf{H} = \mathbf{H}^+ + i\mathbf{H}^-$$

are the complex amplitudes and

$$\mathbf{E} = \left(\mathrm{E}_\rho, \mathrm{E}_\varphi, \mathrm{E}_z\right)^T, \quad \mathbf{H} = \left(\mathrm{H}_\rho, \mathrm{H}_\varphi, \mathrm{H}_z\right)^T,$$

where $(\,\cdot\,)^T$ is the transposition operation. Each field component is a function of three space variables.

We suppose that permittivity inside the waveguide Σ has the form

$$\varepsilon_0\left(\varepsilon_2(\rho) + \alpha\left(\left|\mathrm{E}_\rho\right|^2 + \left|\mathrm{E}_\varphi\right|^2 + |\mathrm{E}_z|^2\right)\right), \tag{3}$$

where $\varepsilon_2(\rho)$ is a sufficiently smooth function and $\alpha > 0$ is a real constant.

It is obvious that $|\mathbf{E}e^{-i\omega t}| = |\mathbf{E}|$. This allows us to rewrite Eq. (2) for the complex amplitudes $\mathbf{E}, \mathbf{H}$:

$$\begin{cases} \mathrm{rot}\mathbf{H} = -i\omega\varepsilon\mathbf{E}, \\ \mathrm{rot}\mathbf{E} = i\omega\mu_0\mathbf{H}. \end{cases} \tag{4}$$

Complex amplitudes of the electromagnetic field $\mathbf{E}, \mathbf{H}$ satisfy the Maxwell Eq. (4), subject to the following boundary conditions: the tangential field components are continuous on the media interface $\rho = r$; the complex amplitudes obey the radiation condition at infinity: the electromagnetic field decays as $O(|\rho|^{-1})$ when $\rho \to \infty$.

The solutions to Eq. (4) are considered in whole space.

Rewrite system (4) in the expanded form:

$$\begin{cases} \frac{1}{\rho}\partial_\varphi \mathrm{E}_z - \partial_z \mathrm{E}_\varphi = i\omega\mu_0 \mathrm{H}_\rho, \\ (\partial_z \mathrm{E}_\rho - \partial_\rho \mathrm{E}_z) = i\omega\mu_0 \mathrm{H}_\varphi, \\ \frac{1}{\rho}\partial_\rho(\rho \mathrm{E}_\varphi) - \frac{1}{\rho}\partial_\varphi \mathrm{E}_\rho = i\omega\mu_0 \mathrm{H}_z, \\ \frac{1}{\rho}\partial_\varphi \mathrm{H}_z - \partial_z \mathrm{H}_\varphi = -i\omega\varepsilon \mathrm{E}_\rho, \\ (\partial_z \mathrm{H}_\rho - \partial_\rho \mathrm{H}_z) = -i\omega\varepsilon \mathrm{E}_\varphi, \\ \frac{1}{\rho}\partial_\rho(\rho \mathrm{H}_\varphi) - \frac{1}{\rho}\partial_\varphi \mathrm{H}_\rho = -i\omega\varepsilon \mathrm{E}_z. \end{cases} \tag{5}$$

The waveguide has circular symmetry; therefore, sought-for solutions are periodical functions with respect to φ. This means that the components of the electromagnetic field have the form

$$\begin{array}{l} \mathrm{E}_\rho = \widetilde{E}_\rho(\rho, z)e^{in\varphi},\ \mathrm{H}_\rho = \widetilde{H}_\rho(\rho, z)e^{in\varphi}, \\ \mathrm{E}_\varphi = \widetilde{E}_\varphi(\rho, z)e^{in\varphi},\ \mathrm{H}_\varphi = \widetilde{H}_\varphi(\rho, z)e^{in\varphi}, \\ \mathrm{E}_\mathrm{z} = \widetilde{E}_z(\rho, z)e^{in\varphi},\ \mathrm{H}_\mathrm{z} = \widetilde{H}_z(\rho, z)e^{in\varphi}. \end{array} \tag{6}$$

where $n = 0, 1, 2, \ldots$

Choosing $n = 0$ in (6) we obtain from (5)

$$\begin{cases} -\partial_z \widetilde{E}_\varphi = i\omega\mu_0 \widetilde{H}_\rho, \\ \frac{1}{\rho}\partial_\rho(\rho \widetilde{E}_\varphi) = i\omega\mu_0 \widetilde{H}_z, \\ \partial_z \widetilde{H}_\rho - \partial_\rho \widetilde{H}_z = -i\omega\varepsilon \widetilde{E}_\varphi, \\ -\partial_z \widetilde{H}_\varphi = -i\omega\varepsilon \widetilde{E}_\rho, \\ \frac{1}{\rho}\partial_\rho(\rho \widetilde{H}_\varphi) = -i\omega\varepsilon \widetilde{E}_z, \\ \partial_z \widetilde{E}_\rho - \partial_\rho \widetilde{E}_z = i\omega\mu_0 \widetilde{H}_\varphi. \end{cases} \tag{7}$$

As it is seen from (7) all equations split into two subsystems. The first and the second subsystems correspond, respectively, to the TE- and TM-waves. However, it is necessary to keep in mind that in (7) TE- and TM-waves are coupled by the Kerr law, which implies that we do not necessarily have pure TE- and TM-waves in the nonlinear media; by this reason we call them hybrid (non-polarized azimuthal-symmetric) waves.

Waves propagating along the boundary depend on z as $e^{i\gamma z}$, where γ is unknown spectral parameter (propagation constant). We will assume that γ is real. It is obvious that $|\mathbf{E}e^{i\gamma z}| = |\mathbf{E}|$ does not depend on z. Thus, the components E_ρ, E_φ, E_z, H_ρ, H_φ, H_z have the following form:

$$\begin{array}{l} \widetilde{E}_\rho \equiv E_\rho(\rho)e^{i\gamma z},\ \widetilde{E}_\varphi \equiv E_\varphi(\rho)e^{i\gamma z},\ \widetilde{E}_z \equiv E_z(\rho)e^{i\gamma z}, \\ \widetilde{H}_\rho \equiv H_\rho(\rho)e^{i\gamma z},\ \widetilde{H}_\varphi \equiv H_\varphi(\rho)e^{i\gamma z},\ \widetilde{H}_z \equiv H_z(\rho)e^{i\gamma z}, \end{array}$$

and the fields $\tilde{\mathbf{E}}$, $\tilde{\mathbf{H}}$ can be represented as

$$\tilde{\mathbf{E}} = \mathrm{Re}\left\{\left(E_\rho e^{i(\gamma z-\omega t)}, E_\varphi e^{i(\gamma z-\omega t)}, E_z e^{i(\gamma z-\omega t)}\right)^T\right\},$$
$$\tilde{\mathbf{H}} = \mathrm{Re}\left\{\left(H_\rho e^{i(\gamma z-\omega t)}, H_\varphi e^{i(\gamma z-\omega t)}, H_z e^{i(\gamma z-\omega t)}\right)^T\right\}.$$

Now from (7) we obtain

$$\begin{cases} i\gamma E_\varphi = -i\omega\mu_0 H_\rho, \\ \frac{1}{\rho}(\rho E_\varphi)' = i\omega\mu_0 H_z, \\ i\gamma H_\rho - H_z' = -i\omega\varepsilon E_\varphi, \\ i\gamma H_\varphi = i\omega\varepsilon E_\rho, \\ \frac{1}{\rho}(\rho H_\varphi)' = -i\omega\varepsilon E_z, \\ i\gamma E_\rho - E_z' = i\omega\mu_0 H_\varphi. \end{cases} \tag{8}$$

where $(\,\cdot\,)' \equiv \partial_\rho$.

After simple transformations from (8) we get

$$\begin{cases} \gamma^2 E_\rho + \gamma(iE_z)' = \omega^2\mu_0\varepsilon E_\rho, \\ \gamma^2 E_\varphi - \left(\rho^{-1}(\rho E_\varphi)'\right)' = \omega^2\mu_0\varepsilon E_\varphi, \\ -\gamma\rho^{-1}(\rho E_\rho)' - \rho^{-1}(\rho(iE_z)')' = \omega^2\mu_0\varepsilon(iE_z), \end{cases} \tag{9}$$

where

$$H_\rho = -\gamma(\omega\mu_0)^{-1}E_\varphi,$$
$$H_z = (i\omega\mu_0)^{-1}\rho^{-1}(\rho E_\varphi)',$$
$$H_\varphi = (i\omega\mu_0)^{-1}(i\gamma E_\rho - E_z').$$

Introduce the notation

$$u_1(\rho) := \mathrm{E}_\rho(\rho),\ u_2(\rho) := \mathrm{E}_\varphi(\rho),\ u_3(\rho) := i\mathrm{E}_z(\rho).$$

From (9) we get

$$\begin{cases} \gamma^2 u_1 + \gamma u_3' = k_0^2\tilde{\varepsilon}u_1, \\ \left(\rho^{-1}(\rho u_2)'\right)' - \gamma^2 u_2 = -k_0^2\tilde{\varepsilon}u_2, \\ \gamma\rho^{-1}(\rho u_1)' + \rho^{-1}(\rho u_3')' = -k_0^2\tilde{\varepsilon}u_3, \end{cases} \tag{10}$$

where $k_0^2 = \omega^2\mu_0\varepsilon_0$ is the wave numbers of free space.

Outside and inside the waveguide $\varepsilon = \widetilde{\varepsilon}\varepsilon_0$, where

$$\widetilde{\varepsilon} = \begin{cases} \varepsilon_2(\rho) + \alpha|\mathbf{u}|^2, & 0 \leqslant \rho \leqslant r, \\ \varepsilon_1, & \rho > r, \end{cases} \tag{11}$$

where $|\mathbf{u}|^2 = u_1^2 + u_2^2 + u_3^2$.

Remark 1 For $\alpha = 0$ system (10) splits into two independent systems, which corresponds to two independent wave propagation problems: for TE and TM guided waves. It is important to realize that in the (10) TE and TM waves are interconnected by the nonlinearity. In other words, we can not say that we have a "clean" TE and TM waves in a nonlinear medium; For this reason, we call them a hybrid waves.

In the domain $\rho > r$ we have $\widetilde{\varepsilon} = \varepsilon_1$. From (10) we obtain the system of linear equations:

$$\begin{cases} \gamma^2 u_1 + \gamma u_3' = k_0^2 \varepsilon_1 u_1, \\ \left(\rho^{-1}(\rho u_2)'\right)' - \gamma^2 u_2 = -k_0^2 \varepsilon_1 u_2, \\ \gamma \rho^{-1}(\rho u_1)' + \rho^{-1}(\rho u_3')' = -k_0^2 \varepsilon_1 u_3. \end{cases}$$

In accordance with the condition at infinity solution of the latter system is

$$\begin{cases} u_1(\rho;\gamma) = \gamma k_1^{-1} C_1 K_1(k_1\rho), \\ u_2(\rho;\gamma) = C_2 K_1(k_1\rho), \\ u_3(\rho;\gamma) = C_1 K_0(k_1\rho), \end{cases} \tag{12}$$

where $k_1^2 = \gamma^2 - k_0^2\varepsilon_1$ and $C_1,\ C_2$ are the constants and functions K_1 and K_0 are the modified Bessel functions (Macdonald functions) [15].

Inside the waveguide $\rho \le r$ we have $\widetilde{\varepsilon} = \varepsilon_2 + \alpha|\mathbf{u}|^2(\rho)$, where

$$|\mathbf{u}|^2 = u_1^2 + u_2^2 + u_3^2, \quad \text{and} \quad (u_1, u_2, u_3)^T.$$

From (10) we obtain the system of nonlinear equations

$$\begin{cases} \gamma^2 u_1 + \gamma u_3' = k_0^2(\varepsilon_2 + \alpha|\mathbf{u}|^2) u_1, \\ \left(\rho^{-1}(\rho u_2)'\right)' - \gamma^2 u_2 = -k_0^2(\varepsilon_2 + \alpha|\mathbf{u}|^2) u_2, \\ \gamma \rho^{-1}(\rho u_1)' + \rho^{-1}(\rho u_3')' = -k_0^2(\varepsilon_2 + \alpha|\mathbf{u}|^2) u_3. \end{cases} \tag{13}$$

Tangential components $E_\varphi, E_z, H_\varphi, H_z$ of the electromagnetic field are continuous at the interface. It follows from the continuity of H_φ that $\tilde{\varepsilon} u_1$ is continuous at the interface. In this way we obtain the transmission conditions for the functions u_1, u_2 and u_3

$$[\tilde{\varepsilon} u_1]|_r = 0,\ [u_2]|_r = 0,\ [u_2']\big|_r = 0,\ [u_3]|_r = 0, \tag{14}$$

where $[f]|_{\rho_0} = \lim\limits_{\rho \to \rho_0 - 0} f(\rho) - \lim\limits_{\rho \to \rho_0 + 0} f(\rho)$.

In propagation problems for polarized TE- and TM-waves, constants C_1 and C_2 can be chosen independently. Nonlinear wave depends on one constant (compare with [16, 17]). In the case of hybrid waves constants C_1 and C_2 are not independent quantities.

We propose that values of functions u_2 and u_3 on the boundary $\rho = r$ can be represented as the following expression

$$u_2(r) = C\cos\theta,\ u_3(r) = C\sin\theta, \tag{15}$$

where constant $C > 0$ is supposed to be known (value of field on the boundary) and parameter $\theta \in [0, 2\pi]$ can be determined using the boundary conditions (14). Thus we obtain the following expression

$$u_2^2(r) + u_3^2(r) = C^2 \cos^2\theta + C^2 \sin^2\theta = C^2 \tag{16}$$

or, using the solutions (12)

$$C_1^2 K_0^2(k_1 r) + C_2^2 K_1^2(k_1 r) = C^2. \tag{17}$$

We choose condition (17) (or (15)) to relate the constants C_1 and C_2, using the representation

$$C_1 = \frac{C \sin\theta}{K_0(k_1 r)} \text{ and } C_2 = \frac{C \cos\theta}{K_1(k_1 r)}.$$

Definition 1 Problem P_H: find real values γ such that for the given constant C there are functions u_1, u_2, and u_3 such that:

1. $|\mathbf{u}|^2 \not\equiv 0$;
2. for $\rho > r$, functions u_1, u_2 and u_3 are determined by the formulas (12);
3. for $\rho < r$ u_1, u_2, and u_3 are the solutions of Eq. (13);
4. functions u_1, u_2 and u_3 satisfy the transmission conditions (14).

Definition 2 The solution γ to the problem P_H are called the eigenvalues, and the corresponding functions u_1, u_2, and u_3 are the eigenfunctions.

3 Numerical Method

For the numerical solution of Problem P_H a method based on the solution to the auxiliary Cauchy problem is proposed (see [9, 18, 19]) which makes it possible in particular to determine and plot the normalized eigenvalues γ with respect to the circular frequency ω. Consider the Cauchy problem for the system of equations

$$\begin{cases} u_1' = -\dfrac{\varepsilon_2' + 2\alpha\left(u_2 u_4 - \frac{1}{\rho}u_2^2 - \gamma u_1 u_3\right)}{\varepsilon_2 + \alpha\left(2u_1^2 + |\mathbf{u}|^2\right)} u_1 - \dfrac{1}{\gamma\rho} \dfrac{\gamma u_1 + \gamma^2 \rho u_3 + 2\alpha\rho k_0^2 u_1^2 u_3}{\varepsilon_2 + \alpha\left(2u_1^2 + |\mathbf{u}|^2\right)} (\varepsilon_2 + \alpha|\mathbf{u}|^2), \\ u_2' = u_4 - \rho^{-1} u_2, \\ u_3' = -\gamma^{-1}\left(\gamma^2 - k_0^2(\varepsilon_2 + \alpha|\mathbf{u}|^2)\right) u_1, \\ u_4' = \left(\gamma^2 - k_0^2(\varepsilon_2 + \alpha|\mathbf{u}|^2)\right) u_2, \end{cases}$$

where $u_4 := u_2' + \dfrac{1}{\rho} u_2$; with the following initial conditions

$$
\begin{aligned}
u_1(r) &:= x,\\
u_2(r) &:= C\cos\theta,\\
u_3(r) &:= C\sin\theta,\\
u_4(r) &:= C\cos\theta\frac{k_1K_1'(k_1r)}{K_1(k_1r)} + \frac{C}{r}\cos\theta,
\end{aligned}
$$

where value x is determined as solution of the following cubic equation

$$ax^3 + bx + c = 0,$$

where

$$
\begin{aligned}
a &= \alpha,\\
b &= \varepsilon_2(r) + \alpha C^2,\\
c &= C\sin\theta\frac{\varepsilon_1\gamma}{k_1}\frac{K_1(k_1r)}{K_0(k_1r)}.
\end{aligned}
$$

To justify the solution technique, we use classical results of the theory of ordinary differential equations concerning the existence and uniqueness of the solution to the Cauchy problem and continuous dependence of the solution on parameters.

In view of the condition of boundedness of the field **E**, **H** in any finite region [1], we obtain the following system of dispersion equations

$$
\begin{cases}
\Delta_E(\gamma) := u_2(0) = 0,\\
\Delta_M(\gamma) := u_1(0) = 0,
\end{cases}
\tag{18}
$$

where quantities $u_1(0)$ and $u_2(0)$ are obtained from the solution to the Cauchy problem.

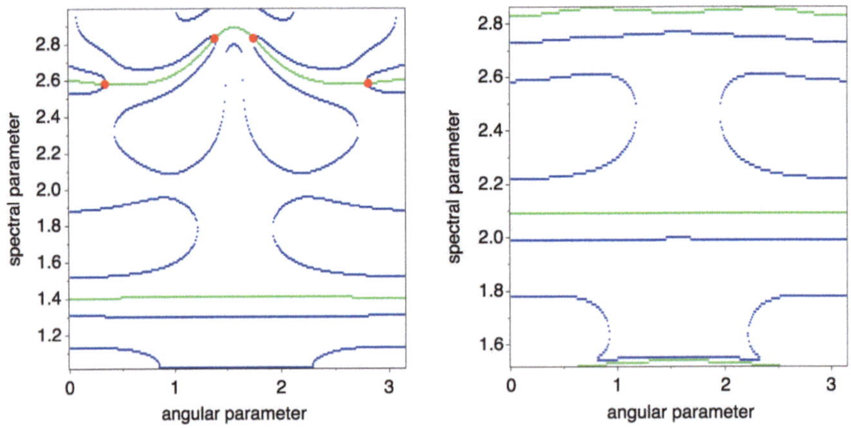

Fig. 2 The graph of spectral parameter γ vs parameter θ for different values of omega ($\omega = 1$ GHz and $\omega = 1.5$ GHz)

Choosing the value of constant C and variate the values of angular parameter θ, we can determine (numerically) the solutions of the equations $\Delta_E = 0$ and $\Delta_M = 0$. This solutions are presented as curves (green color corresponds to the solution of equation $\Delta_E = 0$ and blue to $\Delta_M = 0$) or the dependence of spectral parameter γ versus angular parameter θ. Then we determine points of intersections (red color) of the blue and green curves, these points approximate eigenvalues of Problem P_H (see Fig. 2).

4 Numerical Results

For the numerical solution of Problem P_H a method based on the solution to the auxiliary Cauchy problem is proposed (see [9]) which makes it possible in particular to determine and plot the normalized eigenvalues γ with respect to the circular frequency ω.

The following values of parameters are used for calculations: $C = 1$, $r = 4$, $\varepsilon_1 = 1$, $\varepsilon_2 = 4$, $\alpha = 0.1$.

In Fig. 3 the dependence $\gamma(\omega)$ is plotted. We call $\gamma(\omega)$ the dispersion curve (DC). Red line corresponds to the nonlinear hybrid wave, green and blue curves correspond to the linear TE- and TM-waves, respectively. A grey dashed angle is the domain where the linear problems have a solution. The vertical dotted black line corresponds to the circular frequency $\omega = 1$. Eigenvalues of a particular problem are points of intersections of this dotted line with the DCs.

The behavior of nonlinear DCs is studied with a change in the nonlinearity coefficient α. The constructed curves are indicated in the figure as H, H1, H2. The value of the nonlinearity coefficient is shown in the figure. As follows from the numerical experiment, as the α decreases, the value of the eigenvalue γ increases. This fact shows that when $\alpha \to 0$ nonlinear hybrid modes do not transform into linear TE or TM modes.

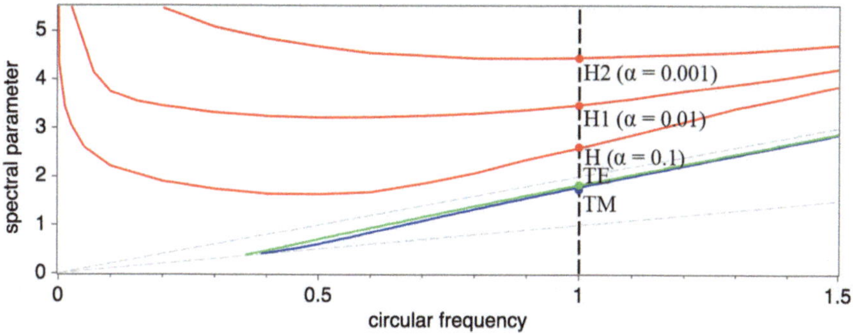

Fig. 3 DCs for linear TE, TM and nonlinear hybrid waves. Marked eigenvalues: linear TE eigenvalue $\gamma \approx 1.819$ (green dot); linear TM eigenvalue $\gamma \approx 1.776$ (blue dot); nonlinear hybrid eigenvalue $\gamma \approx 2.579$ (red dot)

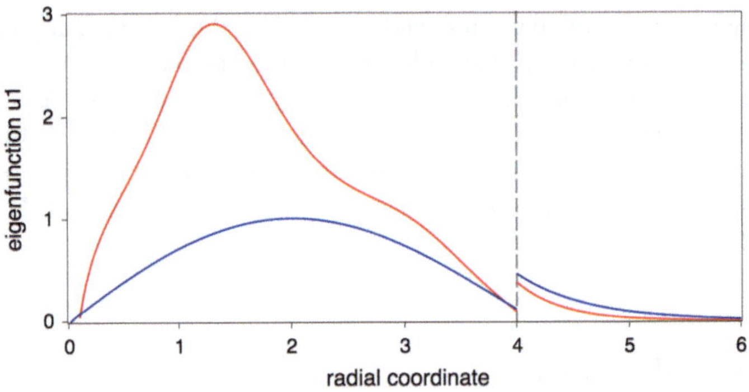

Fig. 4 The graph of eigenfunctions u_1 for linear TM (blue curve) and nonlinear hybrid waves (red curve)

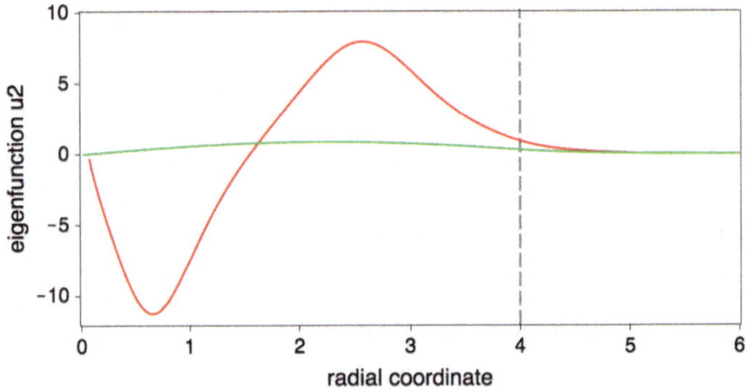

Fig. 5 The graph of eigenfunctions u_2 for linear TE (green curve) and nonlinear hybrid waves (red curve)

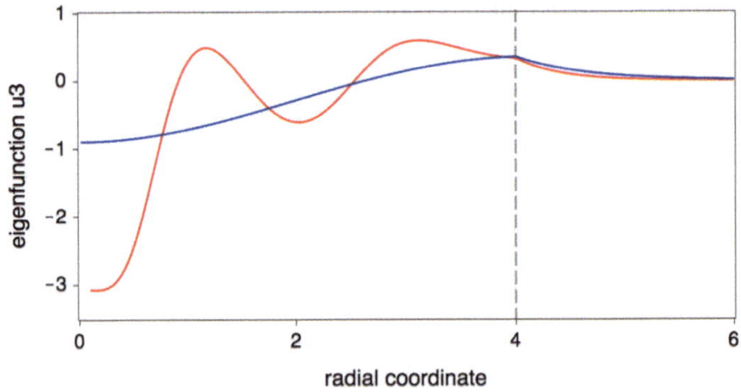

Fig. 6 The graph of eigenfunctions u_3 for linear TM (blue curve) and nonlinear hybrid waves (red curve)

Figs. 4, 5 and 6 display the graphs of eigenfunctions u_1, u_2, and u_3 for the frequency value $\omega = 1$. Color of a curve corresponds to the color of marked eigenvalue in Fig. 3. Graphs of the tangential components of the electromagnetic field consistent are with the physical formulation of the problem; namely, both eigenfunctions u_2 and u_3 are continuous at the interface r, and u_1, u_2, and u_3 decay when $\rho \to \infty$.

5 Conclusion and Discussion

Numerical results which predict the existence of non-polarized azimuthal-symmetric waves in a dielectric waveguide are presented. An interesting fact is that these waves do not correspond to nonlinear TE and TM polarized waves. Natural question that arises is to check experimentally existence of nonlinear hybrid waves.

From the mathematical point of view, analysis of non-polarized azimuthal-symmetric waves can be reduced to a nonlinear setting which may be classified as a new type of nonlinear two-parameter eigenvalue problem. In fact, in the statement considered in this work, there is only one spectral parameter γ while the second, θ, is not the spectral one. General methods to study such problems have not been developed yet. Therefore, the approach set forth in this paper is very interesting for further mathematical investigations

The developed mathematical approaches and numerical techniques can be used for the analysis of the wave propagation in more complicated nonlinear open waveguides.

Acknowledgements This study is supported by the Ministry of Education and Science of the Russian Federation, Project No. 1.894.2017/4.6.

References

1. A.W. Snyder, J. Love, *Optical Waveguide Theory* (Springer, 1983)
2. M.J. Adams, *An Introduction to Optical Waveguides* (Wiley, New York, 1951)
3. P.N. Eleonskii, L.G. Oganesyants, V.P. Silin, Cylindrical nonlinear waveguides. Sov. Phys. JETP **35**, 44–47 (1972)
4. Y.G. Smirnov, D.V. Valovik, Coupled electromagnetic TE-TM wave propagation in a layer with Kerr nonlinearity. J. Math. Phys. **52** (2008)
5. D.V. Valovik, Y.G. Smirnov, E.Y. Smolkin, Nonlinear transmission eigenvalue problem describing TE wave propagation in two-layered cylindrical dielectric waveguides. Comput. Math. Math. Phys. **53**, 1150–1161 (2013)
6. D.V. Valovik, E.Y. Smolkin, Calculation of the propagation constants of inhomogeneous nonlinear double-layer circular cylindrical waveguide by means of the Cauchy problem method. J. Commun. Technol. Electron. **58**, 759–767 (2013)
7. Y.G. Smirnov, E.Y. Smolkin, D.V. Valovik, Nonlinear double-layer Bragg waveguide: analytical and numerical approaches to investigate waveguiding problem. Adv. Numer. Anal. **2014**, 1–11 (2014)

8. E.Y. Smolkin, D.V. Valovik, Guided electromagnetic waves propagating in a two-layer cylindrical dielectric waveguide with inhomogeneous nonlinear permittivity. Adv. Math. Phys. **2015**, 1–11 (2015)
9. E. Smolkin, The Azimuthal symmetric hybrid waves in nonlinear cylindrical waveguide, in *Progress in Electromagnetics Research Symposium Proceedings*, pp. 348–353 (2017)
10. E. Smolkin, Y. Shestopalov, Nonlinear Goubau line: analytical numerical approaches and new propagation regimes. J. Electromagn. Waves Appl. **31**, 781–797 (2017)
11. Y. Smirnov, E. Smolkin, V. Kurseeva, The new type of non-polarized symmetric electromagnetic waves in planar nonlinear waveguide. Appl. Anal. 1–16 (2017)
12. Y. Smirnov, E. Smolkin, On the existence of non-polarized Azimuthal-symmetric electromagnetic waves in circular dielectric waveguide filled with nonlinear isotropic homogeneous medium. Wave Motion **77**, 77–90 (2018)
13. Y. Smirnov, E. Smolkin, Y. Shestopalov, On the existence of non-polarized Azimuthal-symmetric electromagnetic waves in circular metal-dielectric waveguide filled with nonlinear radially inhomogeneous medium. J. Electromagn. Waves Appl. **32**(11), 1389–1408 (2018)
14. J.A. Stretton, *Electromagnetic Theory* (McGraw Hill, New York, 1941)
15. M. Abramowitz, I.A. Stegun, *Handbook of Mathematical Functions* (National Bureau of Standards, Washington, DC, 1972)
16. Y.G. Sminov, H.W. Schürmann, Y.V. Shestopalov, Integral equation approach for the propagation of TE-waves in a nonlinear dielectric cylindrical waveguide. J. Nonlinear Math. Phys. **11**, 9256–268 (2004)
17. H.W. Schürmann, Y.G. Smirnov, Y.V. Shestopalov, Propagation of TE-waves in cylindrical nonlinear dielectric waveguides. Phys. Rev. E **71**(1), 1–10 (2005)
18. E. Smolkin, Goubau line filled with nonlinear medium: numerical study of TM-polarized waves, in *Proceedings of the 2015 International Conference on Electromagnetics in Advanced Applications, ICEAA* (2015), pp. 1–4
19. E. Smolkin, Y.V. Shestopalov, Nonlinear Goubau line: numerical study of TE-polarized waves, in *Progress in Electromagnetics Research Symposium Proceedings* (2015), pp. 1–4

Two-Step Method for Solving Inverse Problem of Diffraction by an Inhomogenous Body

M. Yu. Medvedik, Yu. G. Smirnov and A. A. Tsupak

Abstract The problem of reconstruction of an unknown refractive index $n(x)$ of an inhomogeneous solid P is considered. The refractive index is assumed to be a piecewise-Hölder function. The original boundary value problem for the Helmholtz equation is reduced to the integral Lippmann-Schwinger equation. The incident wave is defined by a point source located outside P. The solution of the inverse problem is obtained in two steps. First, the "current" $J = (k^2 - k_0^2)u$ is determined in the inhomogeneity region. Second, the function $k(x)$ is expressed via the current J and the incident wave u_0. The uniqueness of the solution J to the first-kind integral equation is proved in the class of piecewise-constant functions. The two-step method is verified by solving a test problem with a given refractive index. The comparison between the approximate solutions and the exact one approved the efficiency of the proposed method.

Keywords Refractive index reconstruction · Integral equations · Uniqueness

1 Introduction

The most important methods for solving inverse problems are those that can be numerically implemented. Such methods are mainly based on solving hyperbolic systems of differential equations in the time domain using finite difference or finite element methods, with subsequent minimization of the corresponding functionals and Tikhonov regularization. Such approaches are described in [1–4].

M. Yu. Medvedik · Yu. G. Smirnov · A. A. Tsupak (✉)
Penza State University, 40, Krasnaya street, Penza, Russia
e-mail: altsupak@yandex.ru

M. Yu. Medvedik
e-mail: _medv@mail.ru

Yu. G. Smirnov
e-mail: smirnovyug@mail.ru

L. Beilina and Yu. G. Smirnov (eds.), *Nonlinear and Inverse Problems in Electromagnetics*, Springer Proceedings in Mathematics & Statistics 243,
https://doi.org/10.1007/978-3-319-94060-1_7

In the present paper, we apply the method of integral equations for reconstruction of an unknown function $k(x) = n(x)k_0$ which describes the inhomogeneity of a volume scatterer P. To solve the problem, we use a finite number of values of the given scalar monochromatic field in the near zone, i.e., at several points in some region D, $\overline{P} \cap \overline{D} = \emptyset$. Such a problem arises in microwave tomography.

The original problem for the Helmholtz equation is reduced to the integral Lippmann-Schwinger equation. The equivalency between the original and the integral formulations of the diffraction problem is proved.

The Lippmann-Schwinger equation is convenient to be studied in the $L_2(P)$ space: in this space, the operator of the equation is a Fredholm operator with index zero. In addition, we are eager to use piecewise constant basis functions for numerical implementation of the method.

The solution to the inverse problem is found using the two-step method. First, the "current" $J = (k^2 - k_0^2)u$ is determined in the inhomogeneity domain P. To this end, we use the values of the given total field in the sources' region D. Second, we calculate the function $k(x)$ using the current $J(x)$, the total field $u(x)$ and the incident wave $u_0(x)$ from the Lippmann-Schwinger equation.

2 Theoretical Investigation of the Problem

In this section we study the direct problem of diffraction by a volume body P with a given refractive index.

2.1 Quasi-Classical Formulation of the Direct Scattering Problem

In the homogeneous space $\mathbb{R}^3$ we consider the solid $P = \{x = (x_1, x_2, x_3) : a_i < x_i < b_i\}$, introduce a uniform mesh $\{(x_{1,i_1}, x_{2,i_2}, x_{3,i_3})\}$ in P, and define rectangular subdomains $\Pi_{i_1 i_2 i_3} = \{x : x_{k,i_k} < x_k < x_{k,k+1}\}$, $0 \le i_k \le n-1$.

The inhomogeneity is characterized by the function

$$k(x) = \left\{k_{i_1 i_2 i_3}(x), \quad x \in \Pi_{i_1 i_2 i_3} \right., \tag{1}$$

where $k_{i_1 i_2 i_3}(x) = k_I(x)$ (I denotes multi-index) are Hölder continuous functions.

The medium outside the solid is characterized by a given wavenumber $k_0 > 0$. The incident wave, the scattered field and the total field depend on time harmonically. The incident field $U_0(x, t)$ is defined as follows: $U_0(x, t) = u_0(x)e^{-i\omega t}$, where $u_0(x) = \frac{e^{ik_0|x-x_0|}}{4\pi|x-x_0|}$ and $x_0 \notin \overline{P}$.

Let E_P be a union of all edges of the parallelepipeds Π_I. Give the notation

$$\Pi_I' = \Pi_I \setminus E_P, \quad P' = P \setminus E_P. \tag{2}$$

The direct problem of diffraction is to find a solution $u(x)$ to the following boundary value problem:

$$\begin{aligned}
&(\Delta + k_I^2(x))u(x) = 0, \quad x \in \Pi_I;\\
&(\Delta + k_0^2(x))u(x) = -\delta(x - x_0), \quad x \in \mathbb{R}^3 \setminus \overline{P};\\
&[u]|_{\partial \Pi_I} = 0, \left[\frac{\partial u}{\partial \mathbf{n}}\right]\Bigg|_{\partial \Pi_I'} = 0;\\
&u \in H^1_{loc}(\mathbb{R}^3 \setminus \{x_0\});\\
&\frac{\partial u_s}{\partial r} = ik_0 u_s + o\left(\frac{1}{r}\right).
\end{aligned} \tag{3}$$

We look for quasi-classical solutions of the direct scattering problem that satisfy the conditions

$$u \in C^1(\mathbb{R}^3 \setminus \{x_0\}) \bigcap_I C^2(\Pi_I) \bigcap C^2(\mathbb{R}^3 \setminus (\overline{P} \cup \{x_0\})). \tag{4}$$

2.2 The Lippmann-Schwinger Integral Equation

The original problem (3) can be reduced [5] to the Lippmann-Schwinger equation

$$u(x) - \int_P (k^2(y) - k_0^2) G(x, y) u(y) dy = u_0(x), \quad x \in P. \tag{5}$$

We also use the integral representation of the total field outside the solid P :

$$u(x) = u_0(x) + \int_P (k^2(y) - k_0^2) G(x, y) u(y) dy, \quad x \in \mathbb{R}^3 \setminus (P \cup \{x_0\}). \tag{6}$$

The integral statement of the direct diffraction problem is the system consisting of Eq. (5) and representation (6).

Denote the operator of Eq. (5) by $\mathscr{I} - \mathscr{A} : L_2(P) \to L_2(P)$ Show that a solution $u(x)$ to the problem (5), (6) satisfies the smoothness conditions (4).

Theorem 1 *Let $u \in L_2(P)$ be a solution to (5). Then the smoothness conditions (4) are satisfied by the field $u(x)$ extended outside P by relation (6).*

Proof We have $u_0 \in C^\infty(\mathbb{R}^3 \setminus \{x_0\})$. From the smoothness of the integral operator's kernel at each point $x \notin \overline{P}$ it follows that a solution is infinitely differentiable outside the solid.

Rewrite (5) as follows:

$$u(x) - \int\limits_{\Pi_I} (k_I^2(y) - k_0^2)G(x, y)u(y)dy = \\ = \sum_{J \neq I} \int\limits_{\Pi_J} (k_J^2(y) - k_0^2)G(x, y)u(y)dy + u_0(x), \quad x \in \Pi_I, \tag{7}$$

where the righthand side of the equation is infinitely differentiable in the open domain Π_I.

The inclusion $u \in L_2(P)$ implies $u \in H^2(P)$ and, consequently, $u \in C^\alpha(\overline{P})$ for all $0 < \alpha < 1/2$. Then $\mathscr{A}u \in C^1(\mathbb{R}^3)$ (see [6]). Now we have $u \in C^1(\mathbb{R}^3 \setminus \{x_0\})$ which results also in the energy finiteness condition $u \in H^1_{\text{loc}}(\mathbb{R}^3 \setminus \{x_0\})$.

Let us now prove that $u \in C^2(\Pi_I)$ for any I. We can write according to (7):

$$u(x) = \int\limits_{\Pi_I} (k_I^2(y) - k_0^2)G(x, y)u(y)dy + w(x) = v(x) + w(x), \quad x \in \Pi_I, \tag{8}$$

where $w \in C^\infty(\Pi_I)$.

Let $x_0 \in \Pi_I$ be an arbitrary inner point of the I-th sub-domain such that $d = dist(x_0, \partial \Pi_I) > 0$. Introduce the cut-off function $c \in C^\infty(\Pi_I)$,

$$c(y) = \begin{cases} 1, & y \in B = B_{d/4}(x_0), \\ 0, & y \in B = \Pi_I \setminus B_{3d/4}(x_0). \end{cases}$$

Represent v via the sum

$$v(x) = v_1(x) + v_2(x) = \int\limits_{\Pi_I} (k_I^2(y) - k_0^2)G(x, y)u(y)c(y)dy + \\ + \int\limits_{\Pi_I \setminus B} (k_I^2(y) - k_0^2)G(x, y)u(y)(1 - c(y))dy. \tag{9}$$

Since the smoothness of the kernel in the second term, one obtains $v_2 \in C^\infty(B)$.

Note that $u \in C^{0,\alpha}(\Pi_I)$, $c \in C_0^\infty(\Pi_I)$, and $(k_I^2(y) - k_0^2)G(x, y)u(y)c(y) \in C_0^{0,\alpha}(\mathbb{R}^3)$. From the latter follows the inclusion $v_1 \in C^{2,\alpha}(\mathbb{R}^3)$.

Thus, a solution $u(x)$ is twice differentiable in a vicinity of any point $x_0 \in \Pi_I$, i.e. $u \in C^2(\Pi_I)$. □

From Theorem 1 follows the statement on equivalency between the differential and the integral formulations of the direct scattering problem.

Theorem 2 *If $u(x)$ is a quasi-classical solution to (3) then u satisfies equalities (5) and (6). Vise versa, for any solution $u \in L_2(P)$ to Eq. (5) the total field $u(x)$ extended to $\mathbb{R}^3 \setminus \{x_0\}$ by formula (6) is a quasi-classical solution to the problem (3).*

Proof The former part of the theorem follows from the derivation of the integral equation.

Consider a solution u to Eq. (5) with $u_0 \in C^\infty(\mathbb{R}^3 \setminus \{x_0\})$.

The definition of the term u_1 and smoothness of u_0 in $\mathbb{R}^3 \setminus \{x_0\}$ imply that u is a solution to the Helmholtz equation in the domains Π_I and $\mathbb{R}^3 \setminus (\{x_0\} \cup \overline{P})$.

The scattered field $u_s(x) = \int\limits_P (k^2(y) - k_0^2) G(x, y) u(y) dy$ satisfies the radiation condition. The transmission conditions are fulfilled since the inclusion $u \in C^1(\mathbb{R}^3 \setminus \{x_0\})$. □

Consider the uniqueness theorem.

Theorem 3 *The direct scattering problem has at most one quasi-classical solution.*

Proof Let us show that the corresponding homogeneous boundary value problem (with $u_0 \equiv 0$ in $\mathbb{R}^3$) formulated for the scattered field $u_s \equiv u$ has only the trivial solution $u_s \equiv 0$.

Let $B \supset \overline{P}$ be a sufficiently large ball of radius R. Introduce regions $\Pi_0 := B \setminus (\overline{P})$ with the boundary $\partial \Pi_0 = \partial B \cup \partial P$, and $\Pi_{-1} := \overline{B}^c = \mathbb{R}^3 \setminus \overline{B}$.

Reduce the original problem for the scattered field u_s to a transmission problem in the domains Π_i. Denote the restriction of $u_s(x)$ to $\overline{\Pi}_I$ by $v_I(x)$. The functions v_I are solutions to the Helmholtz equation in the corresponding domains,

$$\begin{aligned} &(\Delta + k_e^2) v_I(x) = 0, \quad x \notin P \ (I = -1, 0), \\ &(\Delta + k_I^2(x)) v_I(x) = 0, \ x \in \Pi_I, \end{aligned} \tag{10}$$

as well as the transmission conditions

$$v_I(x) = v_J(x), \ -\frac{\partial v_I(x)}{\partial \mathbf{n}} = \frac{\partial v_J(x)}{\partial \mathbf{n}}, \quad x \in \partial \Pi_I'. \tag{11}$$

The radiation condition is formulated as follows:

$$\frac{\partial v_{-1}}{\partial r} = i k_0 v_{-1} + o\left(\frac{1}{r}\right). \tag{12}$$

From the first Green formula, the Helmholtz equation and conditions (11) follows the relation (for details, see [5, 7]):

$$\begin{aligned} &\int\limits_{\partial B} \overline{v}_0 v_{0,\mathbf{n}} ds = -\sum_I \int\limits_{\Pi_I} k_I^2 |v_1|^2 dx - k_0^2 \int\limits_{\Pi_0} |v_0|^2 dx + \\ &+ \int\limits_{V_0} |grad v_0|^2 dx + \sum_I \int\limits_{\Pi_I} |grad v_I|^2 dx = -\int\limits_{\partial B} \overline{v}_{-1} \frac{\partial v_{-1}}{\partial \mathbf{n}} ds. \end{aligned} \tag{13}$$

Consider the imaginary part of the latter relation and take into account the radiation conditions:

$$\Im\int\limits_{\partial B}\overline{v}_{-1}\frac{\partial v_{-1}}{\partial \mathbf{n}}ds=\Im\int\limits_{\partial B}(ik_0v_{-1}+o(R^{-1}))\overline{v}_{-1}ds=k_0\int\limits_{\partial B}|v_{-1}|^2ds+o(1)=0. \quad (14)$$

Application of the Rellich lemma (see [8] on p.88) results in equality $v_{-1}(x)\equiv 0$ at each point $x\in \Pi_{-1}$.

Show now that the relation $v_{-1}(x)\equiv 0$ holds at the points of the inhomogeneity domain. Represent the solution u in an external parallelepiped Π_I such that $\partial \Pi_I\cap \partial P=S\neq\emptyset$ as follows:

$$\begin{gathered}u(x)=\int\limits_{P}G(x,y)(k^2(y)-k_0)u(y)dy=\\=\sum_{J\neq I}\int\limits_{\Pi_J}G(x,y)(k_J^2(y)-k_0)u(y)dy+\int\limits_{\Pi_I}G(x,y)(k_I^2(y)-k_0)u(y)dy=\\=v(x)+w(x),\quad x\in P.\end{gathered} \quad (15)$$

It can be shown that the function $u(x)$ is infinitely differentiable in a sufficiently small vicinity U of an arbitrary point $x_0\in S$. Consider the following representation of $w(x)$:

$$\begin{gathered}w(x)=\int\limits_{\Pi_I}G(x,y)(k^2(y)-k_0)u(y)c(y)dy+\\+\int\limits_{\Pi_I\setminus U'}G(x,y)(k^2(y)-k_0)u(y)(1-c(y))dy=w_1(x)+w_2(x),\end{gathered} \quad (16)$$

where $c(y)\in C_0^{0,\alpha}(U)$ is a compactly supported cut-off function such that $c(y)\equiv 1$ in $B_r(x_0)=U'\subset U$. Then the smoothness of the integral operator's kernel implies $w_2\in C^\infty(U')$. Further, the inclusion $w_1\in C^2(\mathbb{R}^3)$ is valid since the term w_1 is a Newtonian potential with the compactly supported smooth density $(k^2(y)-k_0)u(y)c(y)\in C_0^{0,\alpha}(\mathbb{R}^3)$ (see [9] on p. 207).

Thus, the function $u\in C^2(U')$ is a solution to the Helmholtz equation such that $u\equiv 0$ in the sub-domain $U'\setminus P$. From the unique continuity principle ([9], p.212) it follows now that $u\equiv 0$ in the domain U' and, consequently, in the entire parallelepiped Π_I.

Similarly, one can consider all sub-domains Π_I, repeat the above arguments, and conclude that $u\equiv 0$ in P. □

Since $\mathscr{A}$ is a compact operator in $L_2(P)$ then $(\mathscr{I} - \mathscr{A})$ is a Fredholm operator with index zero. From the equivalency theorem it follows that $(\mathscr{I} - \mathscr{A})$ is also an injective operator.

Thus, the operator $(\mathscr{I} - \mathscr{A}) : L_2(P) \to L_2(P)$ is continuously invertible.

2.3 *Statement of the Inverse Problem*

Consider in $\mathbb{R}^3$ an inhomogeneous parallelepiped P characterized by an unknown refractive index $n(x)$. Let $n(x)$ and $k(x) = n(x)k_0$ be piecewise-Hölder functions in the domain P.

Let D be a bounded domain such that $\overline{D} \cap \overline{P} = \emptyset$. Assume that at the points $x \in D$ we are given the known values of the total field $U(x, t) = U_0(x, t) + U_s(x, t)$, $U_s(x, t) = u_s e^{-i\omega t}$ at a fixed frequency ω. The monochromatic incident wave $U_0(x, t)$ is defined as above.

We will reconstruct the function $k(x)$ in P using measurements of the total field $u(x)$ at points of the domain D,

$$\int_P (k^2(y) - k_0^2)G(x, y)u(y)dy = u(x) - u_0(x), \quad x \in D \tag{17}$$

and taking into account the equation

$$u(x) - \int_P (k^2(y) - k_0^2)G(x, y)u(y)dy = u_0(x), \quad x \in P \tag{18}$$

in the inhomogeneity domain P.

2.4 *The Two-Step Method for Solving the Inverse Diffraction Problem*

We define the function $J(x) = (k^2(x) - k_0^2)u(x)$ in the domain P assuming that the condition $|k(x)| \geq \tilde{k} > k_0$ holds everywhere in P. From representation (6) follows the equation

$$\int_P G(x, y)J(y)dy = u(x) - u_0(x) = u_s(x), \quad x \in D \tag{19}$$

for determining the current $J(x)$, whereas Eq. (5) can be rewritten as below:

$$\frac{J(x)}{k^2(x)-k_0^2} - \int_P G(x,y)J(y)dy = u_0(x), \quad x \in P. \tag{20}$$

The idea of the proposed two-step method is as follows:

1. Find the solution J to Eq. (19) in P using the known values of the incident wave $u_0(x)$ and the total field $u(u)$ in the domain D.
2. Calculate the function $k(x)$ inside P using relation (20).

2.5 On Uniqueness of a Solution to the Integral Equation of the First Kind

Note that, in general, a solution to the Fredholm integral equation of the first kind is not unique. It can be shown that the homogeneous integral equation (19) has non-trivial solutions for any k_0. Consider, e.g., an argumentation in the case of the cubic solid $P = [-1; 1]^3 \subset \mathbb{R}^3$.

Define the function $\psi(x) = (1-x_1^2)^2(1-x_2^2)^2(1-x_2^2)^2$ which satisfies the conditions $\psi\big|_{\partial P} = \frac{\partial \psi}{\partial \mathbf{n}}\big|_{\partial P'} = 0$, and define $J(x) = -(\Delta + k_0^2)\psi$. One obtains $\psi(x) = \int\limits_P G(x,y)J(y)dy$.

Introduce the potential $v(x) = \int\limits_P G(x,y)J(y)dy, \quad x \in \mathbb{R}^3$. In P one has $v \equiv \psi$. However, the relation $v \equiv 0$ holds outside the closed cube $\overline{P}$:

$$\begin{aligned} 0 &= \int_{\partial P} \Big(\psi(y)\frac{\partial G(x,y)}{\partial \mathbf{n}} - G(x,y)\frac{\partial \psi(y)}{\partial \mathbf{n}}\Big)ds_y = \\ &= \int_P \Big(\psi(y)\Delta_y G(x,y) - G(x,y)\Delta_y \psi(y)\Big)dy = \\ &= \int_P G(x,y)J(y)dy = v(x), \quad x \notin \overline{P}. \end{aligned} \tag{21}$$

However, one can prove uniqueness of a solution J in the class of piecewise constant functions.

For J, we use the following representation: $J(x) = \sum\limits_I J_I \chi_I(x)$. Here $J_I \in \mathbb{C}$ are the unknown coefficients, and $\chi_I(x)$ are the indicator functions of the domains Π_I,

$$\chi_I(x) = \begin{cases} 1, \; x \in \Pi_I, \\ 0, \; x \notin \Pi_I. \end{cases} \tag{22}$$

Note that the class of piecewise constant solutions is quite sufficient for solving applied problems of physics, medical tomography etc.

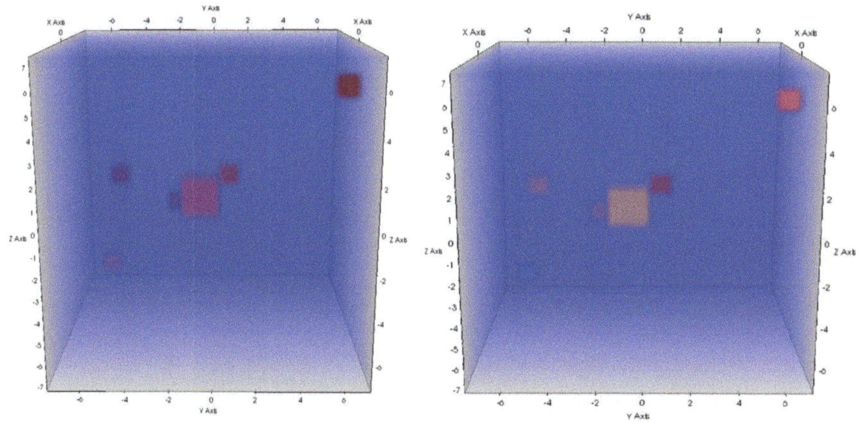

Fig. 1 The real and the imaginary parts of the exact solution $k(x)$

The theorem on uniqueness of a piecewise constant solution $J(x)$ to Eq. (20) is formulated as follows (the proof can be found in [7]).

Theorem 4 *Let $\{\Pi_l\}$ be a fixed set of n^3 rectangular sub-domains in the inhomogeneity domain P. Let $k_0 > \frac{\pi^2 n^3}{2l}$, where $l = \min_i |b_i - a_i|$. If equation*

$$\int_P G(x, y)J(y)dy = u_s(x), \quad x \in D, \ \overline{D} \cap \overline{P} = \emptyset, \ u_s \in C^{\infty}(\overline{D}) \tag{23}$$

has a piecewise constant solution $J(x)$, then this solution is unique.

From Theorem 4 and the equivalency theorem it follows that the solution $k(x)$ to the inverse diffraction problem corresponding to the piecewise constant J is also unique.

3 Numeric Study of the Inverse Problem

Below, we present some results of solving the inverse problem.

Equation (19) is solved using the collocation method. J is represented via linear combination $\sum_{j=1}^{N} c_j \chi_j(x)$ of the indicator functions. The collocation points r_i are in the region D ($i = 1, \ldots, N$). The inhomogeneity domain P is a cube with the edge length of 0.15 m.

The figures show the values of the real and imaginary parts of the function $k(x)$. The representation of the given function $k(x)$ is presented in Fig. 1.

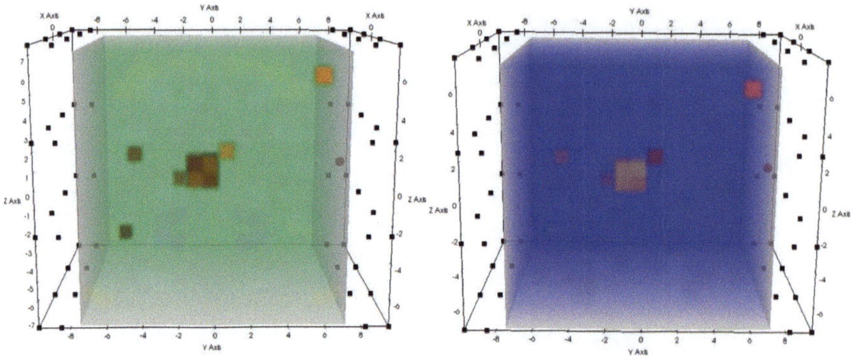

Fig. 2 The real and the imaginary parts of the approximate solution

The solution to the inverse problem is found using the two-step method. The real and imaginary parts of the approximate solution are represented in Fig. 2. The red ball depicts the source of the field, whereas the square dots represent the receivers' location.

Acknowledgements This work was supported by RFBR [the research project No. 18-01-00219].

References

1. A.B. Bakushinsky, M.Yu. Kokurin, *Iterative Methods for Approximate Solution of Inverse Problems* (Springer, New York, 2004)
2. L. Beilina, M. Klibanov, *Approximate Global Convergence and Adaptivity for Coefficient Inverse Problems* (Springer, New York, 2012)
3. S.I. Kabanikhin, A.D. Satybaev, M.A. Shishlenin, *Direct Methods of Solving Multidimensional Inverse Hyperbolic Problems* (VSP, Utrecht, 2004)
4. V.G. Romanov, *Inverse Problems of Mathematical Physics* (VNU, Utrecht, 1986)
5. Yu.G Smirnov, A.A. Tsupak, Method of integral equations in the scalar problem of diffraction on a system consisting of a soft and a hard screen and an inhomogeneous body. Differ. Equ. **50**(9), 1199–1208 (2014)
6. V.S. Vladimirov, *Equations of Mathematical Physics* (Marcel Dekker, New York, 1971)
7. M.Yu. Medvedik, Yu.G. Smirnov, A.A. Tsupak, Inverse problem of diffraction by an inhomogeneous solid with a piecewise hölder refractive index, arXiv:1803.04701
8. D. Colton, R. Kress, *Integral Equation Methods in Scattering Theory* (Wiley-Interscience Publication, New York, 1983)
9. D. Colton, R. Kress, *Inverse Acoustic and Eleectromagnetic Scattering Theory* (Springer-Verlag, Berlin, Heidelberg, 1992)

Numerical Studies of the Lagrangian Approach for Reconstruction of the Conductivity in a Waveguide

L. Beilina and K. Niinimäki

Abstract We consider an inverse problem of reconstructing the conductivity function in a hyperbolic equation using single space-time domain noisy observations of the solution on the backscattering boundary of the computational domain. We formulate our inverse problem as an optimization problem and use Lagrangian approach to minimize the corresponding Tikhonov functional. We present a theorem of a local strong convexity of our functional and derive error estimates between computed and regularized as well as exact solutions of this functional, correspondingly. In numerical simulations we apply domain decomposition finite element-finite difference method for minimization of the Lagrangian. Our computational study shows efficiency of the proposed method in the reconstruction of the conductivity function in three dimensions.

Keywords Coefficient inverse problem · Hyperbolic equation · Tikhonov functional · Lagrangian approach · Domain decomposition

1 Introduction

In this work, we consider the coefficient inverse problem (CIP) of reconstructing the conductivity function in a hyperbolic equation using single observation of the solution of this equation in space and time on the backscattered boundary of the computational domain. In our simulations, backscattered boundary measurements are generated by a single direction of propagation of a plane wave. We solve our CIP via minimization of the corresponding Tikhonov functional and use the Lagrangian approach to minimize

L. Beilina (✉)
Department of Mathematical Sciences, Chalmers University of Technology, University of Gothenburg, 412 96 Gothenburg, Sweden
e-mail: larisa.beilina@chalmers.se

K. Niinimäki
IR4M UMR8081, CNRS, University of Paris-Sud, University of Paris-Saclay, SHFJ, 4 place du Général Leclerc, 91401 Orsay, France
e-mail: kati.niinimaki@u-psud.fr

L. Beilina and Yu. G. Smirnov (eds.), *Nonlinear and Inverse Problems in Electromagnetics*, Springer Proceedings in Mathematics & Statistics 243,
https://doi.org/10.1007/978-3-319-94060-1_8

it. Applying results of [6, 7], we have formulated a theorem of a local strong convexity of this functional in our case and show that the gradient method for minimizing this functional will converge. We have also presented estimates of the norms between computed and regularized solution of the Tikhonov functional via the L_2 norm of the Fréchet derivative of this functional and via the corresponding Lagrangian. In the minimization procedure of the Lagrangian, we applied conjugate gradient method and the domain decomposition finite element/finite difference method of [3]. The method of [3] is convenient for our simulations since it is efficiently implemented in the software package WavES [26] in C++ using PETSc [23] and message passing interface (MPI).

We tested our iterative algorithm by reconstructing a conductivity function that represents some small scatterers as well as smooth function inside the domain of interest. In all of our numerical simulations of this work we induced one non-zero initial condition in the hyperbolic equation accordingly to the theory of the recent work [17]. In [17] it was shown that one non-zero initial condition associated with the observation of the solution of the hyperbolic equation involves uniqueness and stability results in reconstruction of the conductivity function for a cylindrical domains. Our three-dimensional numerical simulations show that we can accurately reconstruct large contrast of the conductivity function as well as its location. In our future work, similar to [5, 8], we are planning to use an adaptive finite element method in order to improve reconstruction of the shapes obtained in this work.

Another method for reconstruction of conductivity function—a layer-stripping algorithm with respect to pseudo-frequency—was presented in [14]. In addition the mathematical model governed by the hyperbolic equation studied in this work can also be considered as a special case of a time-dependent transverse magnetic polarized wave scattering problem or as a simplified acoustic wave model for fluids with variable density and a constant bulk modulus. In recent years, some rapid identification techniques have been developed for solving the elastodynamic inverse problem, for instance, crack/fault identification techniques are developed for cracks having free boundary condition using a reciprocity gap function [12, 13], and linear sampling techniques are designed to locate inclusions in the isotropic elastic medium [10, 20]. To compare performance of the algorithm of this paper with different algorithms of [10, 12–14, 20] can be the subject of a future work.

The paper is organized as follows. In Sect. 2 we formulate the forward and inverse problems. In Sect. 3 we present the Tikhonov functional to be minimized and formulate the theorem of a local strong convexity of this functional. Section 4 is devoted to a Lagrangian approach to solve the inverse problem. In Sect. 5 we present finite element method for the solution of our optimization problem and formulate conjugate gradient algorithm used in computations. Finally, in Sect. 6 we present results of reconstructing the conductivity function in three dimensions.

2 Statement of the Forward and Inverse Problems

Let $\Omega \subset \mathbb{R}^3$ be a convex bounded domain with the boundary $\partial\Omega \in C^3$, $x = (x_1, x_2, x_3) \in \mathbb{R}^3$ and $C^{k+\alpha}$ is Hölder space, $k \geq 0$ is an integer and $\alpha \in (0, 1)$. We use the notation $\Omega_T := \Omega \times (0, T)$, $\partial\Omega_T := \partial\Omega \times (0, T)$, $T > 0$. Next, in our theoretical and numerical investigations we use domain decomposition of the domain Ω into two subregions, Ω_{IN} and Ω_{OUT} such that $\Omega = \Omega_{\mathrm{IN}} \cup \Omega_{\mathrm{OUT}}$, $\partial\Omega_{\mathrm{IN}} \subset \partial\Omega_{\mathrm{OUT}}$, see Fig. 1. The communication between two domains is done through two layers of structured nodes as described in [3]. The boundary $\partial\Omega$ of the domain Ω is such that $\partial\Omega = \partial_1\Omega \cup \partial_2\Omega \cup \partial_3\Omega$ where $\partial_1\Omega$ and $\partial_2\Omega$ are, respectively, front and back sides of the domain Ω. The boundary $\partial_3\Omega$ is the union of the left, right, top and bottom sides of the domain Ω.

We denote by $S_T := \partial_1\Omega \times (0, T)$ the space-time boundary where we will have time-dependent observations of the backscattered field. We use the notation $S_{1,1} := \partial_1\Omega \times (0, t_1]$, $S_{1,2} := \partial_1\Omega \times (t_1, T)$, $S_2 := \partial_2\Omega \times (0, T)$, $S_3 := \partial_3\Omega \times (0, T)$.

Our model problem is as follows

$$\begin{aligned} \frac{\partial^2 u}{\partial t^2} - \nabla \cdot (c \nabla u) &= 0, \text{ in } \Omega_T, \\ u(x, 0) = f_0(x), \quad \partial_t u(x, 0) &= f_1(x) \text{ in } \Omega, \\ \partial_n u &= p(t) \text{ on } S_{1,1}, \\ \partial_n u &= -\partial_t u \text{ on } S_{1,2}, \\ \partial_n u &= -\partial_t u \text{ on } S_2, \\ \partial_n u &= 0 \text{ on } S_3, \end{aligned} \tag{1}$$

which satisfies stability and uniqueness results of [17].

Here, $p(t) \not\equiv 0$ is the incident plane wave generated at the plane $\{x_3 = x_0\}$ and propagating along the x_3-axis. We assume that

$$f_0 \in H^1(\Omega), f_1 \in L_2(\Omega). \tag{2}$$

We assume that in Ω_{OUT} the function $c(x)$ is known and is defined as a constant coefficient $c = 1$. For numerical solution of the problem (1) in Ω_{OUT} we can use either the finite difference or the finite element method. Further in our theoretical considerations we will use the finite element method in both Ω_{OUT} and Ω_{IN}, with known function $c = 1$ in Ω_{OUT} and in the overlapping layer of the structured nodes between Ω_{OUT} and Ω_{IN}. This layer is constructed in a similar way as in [3]. We note that in the numerical simulations of Sect. 6 we use the domain decomposition method of [3] since this method is efficiently implemented in the software package WavES [26] and is convenient for our purposes. We also note that both finite element and finite difference techniques provide the same explicit schemes in Ω_{OUT} in the case of structured mesh in Ω_{OUT}, see [11] for details.

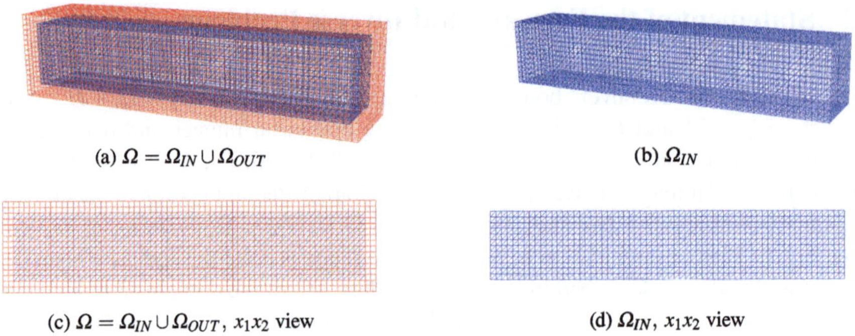

Fig. 1 **a** The hybrid domain $\Omega = \Omega_{IN} \cup \Omega_{OUT}$. Here, wireframe of Ω_{IN} is outlined in blue color and wireframe of Ω_{OUT} in red color. **b** Wireframe of the inner domain Ω_{IN}

We make the following assumptions on the coefficient $c(x)$ in the problem (1):

$$Y = \{c(x) \in [1, d], \; d = const. > 1, \; c(x) = 1 \text{ for } x \in \Omega \diagdown \Omega_{\text{IN}}, \\ c(x) \in C^2(\bar{\Omega})\}. \tag{3}$$

We consider the following

Inverse Problem (IP) *Suppose that the coefficient $c(x)$ of (1) satisfies conditions (3). Assume that the function $c(x)$ is unknown in the domain $\Omega \diagdown \Omega_{\text{OUT}}$. Determine the function $c(x)$ for $x \in \Omega \diagdown \Omega_{\text{OUT}}$, assuming that the following space and time-dependent function $\tilde{u}(x, t)$ is known*

$$u(x, t) = \tilde{u}(x, t), \forall (x, t) \in S_T. \tag{4}$$

From the assumptions (3) it follows that we should know a priori upper and lower bounds of the function $c(x)$. This corresponds to the theory of inverse problems about the availability of a priori information for an ill-posed problem [18, 25]. In applications, the assumption $c(x) = 1$ for $x \in \Omega_{\text{OUT}}$ means that the function $c(x)$ corresponds to the homogeneous domain in $\Omega \diagdown \Omega_{\text{OUT}}$.

3 Tikhonov Functional

We reformulate our inverse problem as an optimization problem and we seek for the function $c(x) \in \Omega \diagdown \Omega_{\text{OUT}}$. This function should fit to the space-time observations $\tilde{u}$ measured at $\partial_1 \Omega$. Thus, we minimize the Tikhonov functional

$$J(c) := J(u, c) = \frac{1}{2} \int_{S_T} (u - \tilde{u})^2 z_\delta(t) d\sigma dt + \frac{1}{2} \gamma \int_{\Omega} (c - c_0)^2 \, dx, \tag{5}$$

where $\tilde{u}$ is the observed field in (4), u satisfies (1), c_0 is the initial guess for c, and γ is the regularization parameter. Here, $z_\delta(t)$ is a cut-off function to impose compatibility conditions at $\overline{\Omega}_T \cap \{t = T\}$ for the adjoint problem (32) which is defined as in [9].

Let us define the L_2 inner product and the norm in Ω_T and Ω, respectively, as

$$
\begin{aligned}
((u, v))_{\Omega_T} &= \int_{\Omega} \int_0^T uv \, dxdt, \\
||u||^2 &= ((u, u))_{\Omega_T}, \\
(u, v)_{\Omega} &= \int_{\Omega} uv \, dx, \\
|u|^2 &= (u, u)_{\Omega}.
\end{aligned}
$$

We also introduce the following spaces of real valued vector functions

$$
\begin{aligned}
H_u^1 &:= \{w \in H^1(\Omega_T) : w(\cdot, 0) = f_0(x), \partial_t w(\cdot, 0) = f_1(x)\}, \\
H_\lambda^1 &:= \{w \in H^1(\Omega_T) : w(\cdot, T) = \partial_t w(\cdot, T) = 0\}, \\
U^1 &= H_u^1(\Omega_T) \times H_\lambda^1(\Omega_T) \times C\left(\overline{\Omega}\right), \\
U^0 &= L_2\left(\Omega_T\right) \times L_2\left(\Omega_T\right) \times L_2\left(\Omega\right).
\end{aligned}
\tag{6}
$$

In our theoretical investigations below we need to reformulate the results of [6, 7] for the case of our **IP**. Below in this section, $|| \cdot ||$ denotes L_2 norm.

We introduce a noise level δ in the function $\tilde{u}(x, t)$ in the Tikhonov functional (5) that corresponds to the theory of ill-posed problems [1, 2, 25];

$$
\tilde{u}(x, t) = u(x, t, c^*) + \tilde{u}_\delta(x, t); \; u(x, t, c^*), \tilde{u}_\delta \in L_2\left(S_T\right), \tag{7}
$$

where $u(x, t, c^*)$ is the exact data corresponding to the exact function c^* in (1), and the function $\tilde{u}_\delta(x, t)$ represents the error in these data. In other words, we can write

$$
\|\tilde{u}_\delta\|_{L_2(S_T)} \leq \delta. \tag{8}
$$

Let $Q_2 = L_2(S_T)$ and Q_1 be the finite dimensional linear space such that

$$
Q_1 = \bigcup_{K_h} \operatorname{span}(V(K_h)), \tag{9}
$$

and

$$
V(K_h) = \{v(x) : v(x) \in H^1(\Omega)\}, \tag{10}
$$

where K_h is the finite-element mesh defined in Sect. 5.

Let $G \subset Q_1$ be a closed bounded convex set satisfying conditions (3). We introduce the operator $F{:}G \to L_2(S_T)$ corresponding to the Tikhonov functional (5) as

$$F(c)(x,t) := u\mid_{S_T} \in L_2(S_T), \tag{11}$$

where $u(x,t,c) := u(x,t)$ is the weak solution of the problem (1) and thus, depends on the function c.

We impose assumption that the operator $F : G \to L_2(S_T)$ is one-to-one. Next, we assume that there exists the exact solution $c^* \in G$ of the equation

$$F\left(c^*\right) = u\left(x,t,c^*\right)\mid_{S_T}. \tag{12}$$

It follows from our assumption that the operator $F : G \to L_2(S_T)$ is one-to-one and thus, for a given function $u(x,t,c^*)$, this solution is unique.

We denote by

$$V_d(c) = \left\{c' \in Q_1 : \left\|c' - c\right\| < d,\ \ \forall d > 0\ \forall c \in Q_1\right\}. \tag{13}$$

We also assume that the operator F has the Lipschitz continuous Frechét derivative $F'(c)$ for $c \in V_1(c^*)$, such that there exist constants $N_1, N_2 > 0$

$$\left\|F'(c)\right\| \leq N_1,\ \left\|F'(c_1) - F'(c_2)\right\| \leq N_2\left\|c_1 - c_2\right\|, \forall c_1, c_2 \in V_1\left(c^*\right). \tag{14}$$

similar to [7] we choose the constant $D = D(N_1, N_2) = const. > 0$ such that

$$\left\|J'(c_1) - J'(c_2)\right\| \leq D\left\|c_1 - c_2\right\|, \forall c_1, c_2 \in V_1(c^*). \tag{15}$$

Through the paper, similar to [7], we assume that

$$\left\|c_0 - c^*\right\| \leq \delta^\xi,\ \xi = const. \in (0,1), \tag{16}$$

$$\gamma = \delta^\zeta, \zeta = const. \in (0, \min(\xi, 2(1-\xi))), \tag{17}$$

where γ is the regularization parameter in (5). Equation (16) means that we assume that initial guess c_0 in (5) is located in a sufficiently small neighborhood $V_{\delta^\xi}(c^*)$ of the exact solution c^*. From Lemma 2.1 and 3.2 of [7] it follows that conditions (16)–(17) ensures that (c^*, c_0) belong to an appropriate neighborhood of the regularized solution of the functional (5).

Below we reformulate Theorem 1.9.1.2 of [6] for the Tikhonov functional (5). Different proofs of this theorem can be found in [6, 21] and in [7] and are straightly applied to our case.

The question of stability and uniqueness of our **IP** is addressed in [17] for the case of the unbounded domain.

Theorem 1. *Let Q_1, Q_2 are two Hilbert spaces such that* $\dim Q_1 < \infty$, $G \subset Q_1$ *is a closed bounded convex set satisfying conditions* (3), $Q_2 = L_2(S_T)$ *and* $F : G \to Q_2$ *is a continuous one-to-one operator.*

Assume that the conditions (7)–(8), (14)–(15) hold. Assume that there exists the exact solution $c^* \in G$ of the equation $F(c^*) = 0$ for the case of the exact data

$u(x,t,c^*)$ in (7). Let the regularization parameter γ in (5) be such that

$$\gamma = \gamma(\delta) = \delta^{2\nu}, \quad \nu = const. \in \left(0, \frac{1}{4}\right), \quad \forall \delta \in (0,1). \tag{18}$$

Let c_0 satisfies the condition (16). Then the Tikhonov functional (5) is strongly convex in the neighborhood $V_\gamma(\delta)(c^*)$ with the strong convexity constant $\alpha = \gamma = \delta^{2\nu}$ such that

$$\|c_1 - c_2\|^2 \le \frac{2}{\delta^{2\nu}} \left(J'(c_1) - J'(c_2), c_1 - c_2\right), \forall c_1, c_2 \in Q_1. \tag{19}$$

Next, there exists the unique regularized solution c_γ of the functional (5) and this solution $c_\gamma \in V_{\delta^{3\nu}/3}(c^*)$. The gradient method of the minimization of the functional (5) which starts at c_0 converges to the regularized solution c_γ of this functional and

$$\left\|c_\gamma - c^*\right\| \le \xi \left\|c_0 - c^*\right\|, \quad \xi \in (0,1). \tag{20}$$

The property (20) means that the regularized solution of the Tikhonov functional (5) provides a better accuracy than the initial guess c_0 if it satisfies condition (16).

The next theorem presents the estimate of the norm $\|c - c_\gamma\|$ via the norm of the Fréchet derivative of the Tikhonov functional (5).

Theorem 2. *Assume that the conditions of Theorem 1 hold. Then for any function $c \in V_{\gamma(\delta)}(c^*)$ the following error estimate is valid*

$$\begin{aligned}\left\|c - c_{\gamma(\delta)}\right\| &\le \frac{2}{\delta^{2\nu}} \left\|P_h J'(c)\right\| \le \frac{2}{\delta^{2\nu}} \left\|J'(c)\right\| \\ &= \frac{2}{\delta^{2\nu}} \left\|L'_c(v(c))\right\| = \frac{2}{\delta^{2\nu}} \left\| \int_0^T (\nabla u(c))(\nabla \lambda(c))\, dt + \gamma(c - c_0) \right\|,\end{aligned} \tag{21}$$

where $c_{\gamma(\delta)}$ is the minimizer of the Tikhonov functional (5) computed with the regularization parameter γ and $P_h : L_2(\Omega) \to Q_1$ is the operator of orthogonal projection of the space $L_2(\Omega)$ on its subspace Q_1, $L'_c(u(c))$ is the Fréchet derivative of the Lagrangian (26) given by (34).

Proof. Since $c_\gamma := c_{\gamma(\delta)}$ is the minimizer of the functional (5) on the set G and $c_\gamma \in Int\,(G)$, then $P_h J'(c_\gamma) = 0$, or using (34) we can write

$$P_h J'_c(c_\gamma) = 0. \tag{22}$$

similar to Theorem 4.11.2 of [6], since $c - c_\gamma \in Q_1$, then

$$(J'(c) \; - J'(c_\gamma), c - c_\gamma) = (P_h J'(c) - P_h J'(c_\gamma), c - c_\gamma).$$

Hence, using (22) and the strong convexity property (19) we can write that

$$\begin{aligned}\left\|c-c_\gamma\right\|^2 &\le \frac{2}{\delta^{2\nu}}\left(J'(c)-J'(c_\gamma),c-c_\gamma\right)\\ &=\frac{2}{\delta^{2\nu}}\left(P_hJ'(c)-P_hJ'(c_\gamma),c-c_\gamma\right)\\ &=\frac{2}{\delta^{2\nu}}(P_hJ'(c),c-c_\gamma)\\ &\le\frac{2}{\delta^{2\nu}}\left\|P_hJ'(c)\right\|\cdot\left\|c-c_\gamma\right\|.\end{aligned}$$

Thus, from the expression above we get

$$\left\|c-c_\gamma\right\|^2\le\frac{2}{\delta^{2\nu}}\left\|P_hJ'(c)\right\|\cdot\left\|c-c_\gamma\right\|. \tag{23}$$

Using the fact

$$\left\|P_hJ'(c)\right\|_{L_2(\Omega)}\le\left\|J'(c)\right\|_{L_2(\Omega)}$$

together with (34) and (35) and dividing the expression (23) by $\left\|c-c_\gamma\right\|$, we obtain the inequality (21).

In our final theorem we present the error between the computed and exact solutions of the functional (5).

Theorem 3. *Assume that the conditions of Theorem 1 hold. Then for any function* $c\in V_{\gamma(\delta)}(c^*)$ *the following error estimate holds*

$$\left\|c-c^*\right\|\le\frac{2}{\delta^{2\nu}}\left\|\int_0^T(\nabla u(c))(\nabla\lambda(c))\,dt+\gamma(c-c_0)\right\|+\xi\left\|c_0-c^*\right\|. \tag{24}$$

Proof. Applying Theorem 2 and inequality (20) we get the inequality (24)

$$\begin{aligned}\left\|c-c^*\right\| &=\left\|c-c_{\gamma(\delta)}+c_{\gamma(\delta)}-c^*\right\|\le\left\|c-c_{\gamma(\delta)}\right\|+\left\|c_{\gamma(\delta)}-c^*\right\|\\ &\le\frac{2}{\delta^{2\nu}}\left\|\int_0^T(\nabla u(c))(\nabla\lambda(c))\,dt+\gamma(c-c_0)\right\|+\xi\left\|c_0-c^*\right\|. \end{aligned}\tag{25}$$

4 Lagrangian Approach

In this section, we will present the Lagrangian approach to solve the inverse problem **IP**. To minimize the Tikhonov functional (5) we introduce the Lagrangian

$$\begin{aligned}L(v) &=J(u,c)-\int_{\Omega_T}\frac{\partial\lambda}{\partial t}\frac{\partial u}{\partial t}\,dxdt+\int_{\Omega_T}(c\nabla u)(\nabla\lambda)\,dxdt\\ &-\int_\Omega\lambda(x,0)f_1(x)\,dx-\int_{S_{1,1}}\lambda p(t)\,d\sigma dt+\int_{S_{1,2}}\lambda\partial_t u\,d\sigma dt+\int_{S_2}\lambda\partial_t u\,d\sigma dt,\end{aligned}\tag{26}$$

where $v = (u, \lambda, c) \in U^1$. We search for a stationary point of (26) with respect to v satisfying $\forall \bar{v} = (\bar{u}, \bar{\lambda}, \bar{c}) \in U^1$

$$L'(v; \bar{v}) = 0, \tag{27}$$

where $L'(v; \cdot)$ is the Jacobian of L at v. We can rewrite the equation (27) as

$$L'(v; \bar{v}) = \frac{\partial L}{\partial \lambda}(v)(\bar{\lambda}) + \frac{\partial L}{\partial u}(v)(\bar{u}) + \frac{\partial L}{\partial c}(v)(\bar{c}) = 0. \tag{28}$$

To find the Frechét derivative (27) of the Lagrangian (26) we consider $L(v + \bar{v}) - L(v)$ $\forall \bar{v} \in U^1$. Then we single out the linear part of the obtained expression with respect to $\bar{v}$. When we derive the Frechét derivative we assume that in the Lagrangian (26) functions in $v = (u, \lambda, c) \in U^1$ can be varied independently on each other. We assume that $\lambda\,(x, T) = \partial_t \lambda\,(x, T) = 0$ and seek to impose conditions on the function λ such that $L\,(u, \lambda, c) := L\,(v) = J\,(u, c)$. Next, we use the fact that $\lambda(x, T) = \frac{\partial \lambda}{\partial t}(x, T) = 0$ and $u(x, 0) = f_0(x)$, $\frac{\partial u}{\partial t}(x, 0) = f_1(x)$, as well as $c = 1$ on $\partial \Omega$, together with boundary conditions $\partial_n u = 0$ and $\partial_n \lambda = 0$ on S_3. The Eq. (27) expresses that for all $\bar{u}$,

$$\begin{aligned} 0 = \frac{\partial L}{\partial \lambda}(u)(\bar{\lambda}) = & -\int_{\Omega_T} \frac{\partial \bar{\lambda}}{\partial t}\frac{\partial u}{\partial t}\, dxdt + \int_{\Omega_T} (c\nabla u)(\nabla \bar{\lambda})\, dxdt - \int_{\Omega} \bar{\lambda}(x, 0) f_1(x)\, dx \\ & - \int_{S_{1,1}} \bar{\lambda} p(t)\, d\sigma dt + \int_{S_{1,2}} \bar{\lambda} \partial_t u\, d\sigma dt + \int_{S_2} \bar{\lambda} \partial_t u\, d\sigma dt, \\ & \forall \bar{\lambda} \in H^1_{\lambda}(\Omega_T), \end{aligned} \tag{29}$$

$$\begin{aligned} 0 = \frac{\partial L}{\partial u}(u)(\bar{u}) & = \int_{S_T} (u - \tilde{u})\, \bar{u}\, z_{\delta}\, d\sigma dt - \int_{\Omega} \frac{\partial \lambda}{\partial t}(x, 0)\bar{u}(x, 0) dx - \int_{S_{1,2} \cup S_2} \frac{\partial \lambda}{\partial t} \bar{u}\, d\sigma dt \\ & - \int_{\Omega_T} \frac{\partial \lambda}{\partial t}\frac{\partial \bar{u}}{\partial t}\, dxdt + \int_{\Omega_T} (c\nabla \lambda)(\nabla \bar{u})\, dxdt, \quad \forall \bar{u} \in H^1_u(\Omega_T). \end{aligned} \tag{30}$$

Finally, we obtain the equation which expresses stationarity of the gradient with respect to c:

$$0 = \frac{\partial L}{\partial c}(u)(\bar{c}) = \int_{\Omega_T} (\nabla u)(\nabla \lambda)\bar{c}\, dxdt + \gamma \int_{\Omega} (c - c_0)\bar{c}\, dx, \ x \in \Omega. \tag{31}$$

The Eq. (29) is the weak formulation of the state equation (1) and the Eq. (30) is the weak formulation of the following adjoint problem

$$
\begin{aligned}
\frac{\partial^2 \lambda}{\partial t^2} - \nabla \cdot (c \nabla \lambda) &= -(u - \tilde{u})|_{S_T} z_\delta \text{ in } \Omega_T, \\
\lambda(\cdot, T) &= \frac{\partial \lambda}{\partial t}(\cdot, T) = 0, \\
\partial_n \lambda &= \partial_t \lambda, \text{ on } S_{1,2}, \\
\partial_n \lambda &= \partial_t \lambda, \text{ on } S_2, \\
\partial_n \lambda &= 0, \text{ on } S_3.
\end{aligned} \tag{32}
$$

We note that we have positive sign here in absorbing boundary conditions. However, after discretization in time of these conditions we will obtain the same schemes for computation of λ^{k-1} as for the computation of u^{k+1} in the forward problem since we solve the adjoint problem backward in time.

Let now the functions $u(c)$, $\lambda(c)$ be the exact solutions of the forward and adjoint problems, respectively, for the known function c satisfying condition (20). Then with $v(c) = (u(c), \lambda(c), c) \in U^1$ and using the fact that for exact solutions $u(c)$, $\lambda(c)$ from (26) we have

$$
J(u(c), c) = L(v(c)) \tag{33}
$$

and assuming that solutions $u(c)$, $\lambda(c)$ are sufficiently stable (see Chap. 5 of book [22] for details), we can write that the Frechét derivative of the Tikhonov functional is given by

$$
J'(c) := J'(u(c), c) = \frac{\partial J}{\partial c}(u(c), c) = \frac{\partial L}{\partial c}(v(c)). \tag{34}
$$

Inserting (31) into (34) we get

$$
J'(c)(x) := J'(u(c), c)(x) = \int_0^T (\nabla u(c))(\nabla \lambda(c))(x, t)\, dt + \gamma (c - c_0)(x). \tag{35}
$$

We note that the Lagrangian (26) and the optimality conditions (29), (30) will be the same, when the homogeneous initial conditions are used in the model problem (1), and only the terms containing the initial conditions will disappear.

5 Finite Element Method for the Solution of an Optimization Problem

In this section, we formulate the finite element method for the solution of the forward problem (1) and the adjoint problem (32). We also present a conjugate gradient method for the solution of our **IP**.

5.1 Finite Element Discretization

We discretize $\Omega_{FEM} \times (0, T)$ denoting by $K_h = \{K\}$ the partition of the domain Ω_{FEM} into tetrahedra K ($h = h(x)$ being a mesh function, defined as $h|_K = h_K$, representing the local diameter of the elements), and we let J_τ be a partition of the time interval $(0, T)$ into time sub-intervals $J = (t_{k-1}, t_k]$ of uniform length $\tau = t_k - t_{k-1}$. We assume also a minimal angle condition on the K_h [11].

To formulate the finite element method, we define the finite element spaces C_h, W_h^u and W_h^λ. First we introduce the finite element trial space W_h^u for u defined by

$$W_h^u := \{w \in H_u^1 : w|_{K\times J} \in P_1(K) \times P_1(J), \forall K \in K_h, \forall J \in J_\tau\},$$

where $P_1(K)$ and $P_1(J)$ denote the set of piecewise-linear functions on K and J, respectively. We also introduce the finite element test space W_h^λ defined by

$$W_h^\lambda := \{w \in H_\lambda^1 : w|_{K\times J} \in P_1(K) \times P_1(J), \forall K \in K_h, \forall J \in J_\tau\}.$$

To approximate function $c(x)$ we will use the space of piecewise constant functions $C_h \subset L_2(\Omega)$,

$$C_h := \{u \in L_2(\Omega) : u|_K \in P_0(K), \forall K \in K_h\}, \tag{36}$$

where $P_0(K)$ is the piecewise constant function on K.

Next, we define $V_h = W_h^u \times W_h^\lambda \times C_h$. Usually $\dim V_h < \infty$ and $V_h \subset U^1$ as a set and we consider V_h as a discrete analogue of the space U^1. We introduce the same norm in V_h as the one in U^0, $\|\bullet\|_{V_h} := \|\bullet\|_{U^0}$, from which it follows that in finite dimensional spaces all norms are equivalent and in our computations we compute coefficients in the space C_h. The finite element method now reads: Find $v_h \in V_h$, such that

$$L'(v_h)(\bar{v}) = 0 \; \forall \bar{v} \in V_h. \tag{37}$$

Using (37) we can write the finite element method for the forward problem (1) (for convenience we will use here and in Sect. 5.2 $f_0 = f_1 = 0$ in Ω_T): Find $u_h \in W_h^u$, such that $\forall \bar{\lambda} \in W_h^\lambda$ and for known $c_h \in C_h$,

$$\begin{aligned} &-\int_{\Omega_T} \frac{\partial \bar{\lambda}}{\partial t}\frac{\partial u_h}{\partial t}\, dxdt - \int_{S_{1,1}} p(t)\bar{\lambda}\, d\sigma dt \\ &+\int_{S_{1,2}\cup S_2} \partial_t u_h \bar{\lambda}\, d\sigma dt + \int_{\Omega_T} (c_h \nabla u_h)(\nabla \bar{\lambda})\, dxdt = 0. \end{aligned} \tag{38}$$

Similarly, the finite element method for the adjoint problem (32) in Ω_T reads: Find $\lambda_h \in W_h^\lambda$, such that $\forall \bar{u} \in W_h^u$ and for known $u_h \in W_h^u$, $c_h \in C_h$,

$$
-\int_{\Omega_T} \frac{\partial \lambda_h}{\partial t}\frac{\partial \bar{u}}{\partial t}\,dxdt + \int_{S_T} (u_h - \tilde{u}) z_\sigma \bar{\lambda}\, d\sigma dt
$$

$$
-\int_{S_{1,2}\cup S_2} \partial_t \lambda_h \bar{u}\, d\sigma dt + \int_{\Omega_T} (c_h \nabla \lambda_h)(\nabla \bar{u})\, dxdt = 0. \tag{39}
$$

5.2 Fully Discrete Scheme

We expand functions $u_h(x,t)$ and $\lambda_h(x,t)$ in terms of the standard continuous piecewise linear functions $\{\varphi_i(x)\}_{i=1}^M$ in space and $\{\psi_k(t)\}_{k=1}^N$ in time, substitute them into (38) and (39), and compute explicitly all time integrals which will appear in the system of discrete equations. Finally, we obtain the following system of linear equations for the forward and adjoint problems (1), (32), correspondingly (for convenience we consider here $f_0 = f_1 = 0$):

$$
\begin{aligned}
M(\mathbf{u}^{k+1} - 2\mathbf{u}^k + \mathbf{u}^{k-1}) &= \tau^2 G^k - \tau^2 K \mathbf{u}^k - \frac{1}{2}\tau M_{\partial\Omega}(\mathbf{u}^{k+1} - \mathbf{u}^{k-1}),\\
M(\boldsymbol{\lambda}^{k+1} - 2\boldsymbol{\lambda}^k + \boldsymbol{\lambda}^{k-1}) &= -\tau^2 S^k - \tau^2 K \boldsymbol{\lambda}^k + \frac{1}{2}\tau M_{\partial\Omega}(\boldsymbol{\lambda}^{k+1} - \boldsymbol{\lambda}^{k-1}),
\end{aligned} \tag{40}
$$

with initial conditions :

$$
u(\cdot,0) = \frac{\partial u}{\partial t}(\cdot,0) = 0, \tag{41}
$$

$$
\lambda(\cdot,T) = \frac{\partial \lambda}{\partial t}(\cdot,T) = 0. \tag{42}
$$

Here, M and $M_{\partial\Omega}$ are the block mass matrix in space and mass matrix at the boundary $\partial\Omega$, respectively, K is the block stiffness matrix, G^k and S^k are load vectors at time level t_k, $\mathbf{u}^k$ and $\boldsymbol{\lambda}^k$ denote the nodal values of $u_h(\cdot,t_k)$ and $\lambda_h(\cdot,t_k)$, respectively and τ is a time step. For details of obtaining this system of discrete equations and computing the time integrals in it, as well as for obtaining then the system (40), we refer to [3].

Let us define the mapping F_K for the reference element $\hat{K}$ such that $F_K(\hat{K}) = K$ and let $\hat{\varphi}$ be the piecewise linear local basis function on the reference element $\hat{K}$ such that $\varphi \circ F_K = \hat{\varphi}$. Then the explicit formulas for the entries in system (40) at each element K can be given as:

$$
\begin{aligned}
M_{i,j}^K &= (\varphi_i \circ F_K, \varphi_j \circ F_K)_K,\\
K_{i,j}^K &= (c_i \nabla\varphi_i \circ F_K, \nabla\varphi_j \circ F_K)_K,\\
G_j^K &= (p^k, \varphi_j \circ F_K)_{K\in S_{1,1}},\\
S_j^K &= ((u_{h_{i,k}} - \tilde{u}_{i,k})|_{\partial_1\Omega} z_\delta, \varphi_j \circ F_K)_K,
\end{aligned} \tag{43}
$$

where $(\cdot,\cdot)_K$ denotes the $L_2(K)$ scalar product and ∂K is the part of the boundary of element K which lies at $\partial\Omega_{FEM}$. Here, $u_{h_{i,k}} = u(x_i, t_k)$ are computed solutions of the forward problem (1), and $\tilde{u}_{i,k} = \tilde{u}(x_i, t_k)$ are discrete measured values of $\tilde{u}(x,t)$ at S_T at the point $x_i \in K_h$ and time moment $t_k \in J_k$.

To obtain an explicit scheme we approximate M with the lumped mass matrix M^L (for further details, see [15]). Next, we multiply (40) by $(M^L)^{-1}$ and get the following explicit method inside Ω_{FEM}:

$$\begin{aligned} \mathbf{u}^{k+1} &= \tau^2 (M^L)^{-1} G^k + (2 - \tau^2 (M^L)^{-1} K)\mathbf{u}^k - \mathbf{u}^{k-1}, \\ \boldsymbol{\lambda}^{k-1} &= -\tau^2 (M^L)^{-1} S^k + (2 - \tau^2 (M^L)^{-1} K)\boldsymbol{\lambda}^k - \boldsymbol{\lambda}^{k+1}. \end{aligned} \tag{44}$$

In the formulas above the terms with $M_{\partial\Omega}$ disappeared since we used schemes (44) only inside Ω_{FEM}.

Finally, for reconstructing $c(x)$ we can use a gradient-based method with an appropriate initial guess values of c_0 which satisfies the condition (16). We have the following expression for the discrete version of the gradient with respect to coefficient c in (31):

$$g_h(x) = \int_0^T \nabla u_h \nabla \lambda_h dt + \gamma(c_h - c_0). \tag{45}$$

Here, λ_h and u_h are computed values of the adjoint and forward problems, respectively, using explicit schemes (44), and c_h is approximated value of the computed coefficient.

5.3 *The Algorithm*

We use conjugate gradient method for the iterative update of approximations c_h^m of the function c_h, where m is the number of iteration in our optimization procedure. We denote

$$g^m(x) = \int_0^T \nabla u_h^m \nabla \lambda_h^m dt + \gamma(c_h^m - c_0), \tag{46}$$

where functions $u_h\left(x,t,c_h^m\right)$, $\lambda_h\left(x,t,c_h^m\right)$ are computed by solving the state and the adjoint problems with $c := c_h^m$.

Algorithm

Step 0. Choose a mesh K_h in Ω and a time partition J of the time interval $(0,T)$. Start with the initial approximation $c_h^0 = c_0$ and compute the sequences of c_h^m via the following steps:

Step 1. Compute solutions $u_h\left(x,t,c_h^m\right)$ and $\lambda_h\left(x,t,c_h^m\right)$ of the state (1) and the adjoint (32) problems on K_h and J using explicit schemes (44).

Step 2. Update the coefficient $c_h := c_h^{m+1}$ on K_h and J using the conjugate gradient method

$$c_h^{m+1} = c_h^m + \alpha^m d^m(x),$$

where α is the step-size in the gradient update [24] which is computed as

$$\alpha^m = \frac{((g^m, d^m))}{\gamma \|d^m\|^2},$$

and

$$d^m(x) = -g^m(x) + \beta^m d^{m-1}(x),$$

with

$$\beta^m = \frac{||g^m(x)||^2}{||g^{m-1}(x)||^2},$$

where $d^0(x) = -g^0(x)$.

Step 3. Stop computing c_h^m and obtain the function c_h if either $||g^m||_{L_2(\Omega)} \leq \theta$ or norms $||g^m||_{L_2(\Omega)}$ are stabilized. Here, θ is the tolerance in updates m of the gradient method. Otherwise set $m := m + 1$ and go to step 1.

6 Numerical Studies

In this section, we present numerical simulations of the reconstruction of unknown function $c(x)$ of the Eq. (1) inside a domain Ω_{FEM} using the algorithm of Sect. 5.3.

For computations of the numerical approximations u_h of the forward and λ_h of the adjoint problems in step 1 of the algorithm of Sect. 5.3, we use the domain decomposition method of [3]. We decompose Ω into two subregions Ω_{IN} and Ω_{OUT} as described in Sect. 2, and we define $\Omega_{FEM} := \Omega_{IN}$ and $\Omega_{FDM} := \Omega_{OUT}$ such that $\Omega = \Omega_{FEM} \cup \Omega_{FDM}$. In Ω_{FEM} we use finite elements as described in Sect. 5.1. In Ω_{FDM} we will use finite difference method. The boundary $\partial\Omega$ is such that $\partial\Omega = \partial_1\Omega \cup \partial_2\Omega \cup \partial_3\Omega$, see Sect. 2 for description of this boundary.

We assume that the conductivity function $c(x)$ is known inside Ω_{FDM} and we set it to be $c(x) = 1$. The goal of our numerical tests is to reconstruct small inclusions with $c = 4.0$ inside every small scatterer, which can represent defects inside a waveguide. We also test our reconstruction algorithm when $c(x)$ represents a smooth function. We consider four different case studies with different geometries of the scatterers:

(i) 3 scatterers of different size located on the same plane with respect to the wave propagation;
(ii) 3 scatterers of different size non-uniformly located inside the waveguide;
(iii) $c(x)$ is smooth function which is presented by one spike of Gaussian function;
(iv) $c(x)$ is smooth function presented by three spikes of Gaussian functions.

Figures 2 and 8 present the considered geometries of the case studies.

In [3] it was shown that the best reconstruction results for our set-ups are obtained for the wave length λ with the frequency $\omega = 40$ in the initialization of a plane wave in (48). Thus, for all test cases (i)–(iv) we choose $\omega = 40$ in (48) and solve the model problem (1) with non-homogeneous initial condition $f_0(x)$ and with $f_1(x) = 0$ in (1). In all our tests we initialized initial conditions at backscattered side $\partial_1 \Omega$ as

$$\begin{aligned} u(x,0) &= f_0(x) = \exp^{-(x_1^2+x_2^2+x_3^3)} \cdot \cos t|_{t=0} = \exp^{-(x_1^2+x_2^2+x_3^3)}, \\ \frac{\partial u}{\partial t}(x,0) &= f_1(x) = -\exp^{-(x_1^2+x_2^2+x_3^3)} \cdot \sin t|_{t=0} \equiv 0. \end{aligned} \tag{47}$$

The domain decomposition is done in the same way, as described above, for all of the case studies. Next, we introduce dimensionless spatial variables $x' = x/(1m)$ such that the domain Ω_{FEM} is transformed into dimensionless computational domain

$$\Omega_{FEM} = \{x = (x_1, x_2, x_3);\ x_1 \in (-3.2, 3.2),\ x_2 \in (-0.6, 0.6),\ x_3 \in (-0.6, 0.6)\}.$$

The dimensionless size of our computational domain Ω for the forward problem is

$$\Omega = \{x = (x_1, x_2, x_3);\ x_1 \in (-3.4, 3.4),\ x_2 \in (-0.8, 0.8),\ x_3 \in (-0.8, 0.8)\}.$$

The space mesh in Ω_{FEM} and in Ω_{FDM} consists of tetrahedral and cubes, respectively. We choose the mesh size $h = 0.1$ in our geometries in the domain decomposition FEM/FDM method, as well as in the overlapping regions between Ω_{FEM} and Ω_{FDM}.

We generate backscattered measurements $\tilde{u}$ at S_T in Ω by a single plane wave $p(t)$ initialized at $\partial_1 \Omega$ in time $T = [0, 3.0]$ such that

$$p(t) = \begin{cases} \sin(\omega t), & \text{if } t \in \left(0, \frac{2\pi}{\omega}\right), \\ 0, & \text{if } t > \frac{2\pi}{\omega}. \end{cases} \tag{48}$$

For the generation of the simulated backscattered data for cases (i)–(ii) we first define exact function $c(x) = 4$ inside small scatterers, and $c(x) = 1$ at all other points of the computational domain Ω_{FEM}. The function $c(x)$ for cases (iii)–(iv) is defined in Sects. 6.3 and 6.4, respectively. Next, we solve the forward problem (1) on a locally refined mesh in Ω_{FEM} in time $T = [0, 3.0]$ with a plane wave as in (48). This allows us to avoid problem with variational crimes. Since we apply explicit schemes (40) in our computations, we use the time step $\tau = 0.006$ which satisfies the CFL condition, see details in [3, 16].

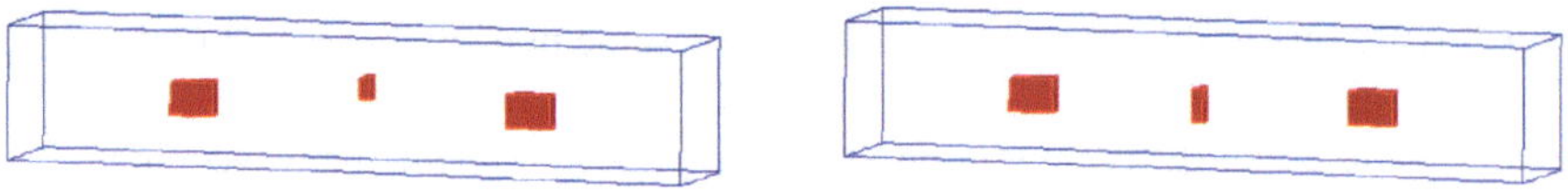

(a) Test (i) (b) Test (ii)

Fig. 2 Exact values of the conductivity function in cases (i) and (ii) are: $c = 4.0$ inside all small scatterers of **a**–**b**, and $c = 1.0$ everywhere else in Ω_{FEM}

For all case studies, we start the optimization algorithm with guess values of the parameter $c(x) = 1.0$ at all points in Ω. Such choice of the initial guess provides a good reconstruction for functions $c(x)$ and corresponds to starting the gradient algorithm from the homogeneous domain, see also [3, 5, 9] for a similar choice of initial guess. In tests (i)–(ii) the minimal and maximal values of the functions $c(x)$ in our computations belongs to the following set of admissible parameters

$$M_c \in \{c \in C(\overline{\Omega}) | 1 \leq c(x) \leq 5\}. \tag{49}$$

We regularize the solution of the inverse problem by starting computations with regularization parameter $\gamma = 0.01$ in (5) which satisfies the condition (17). Our computational studies have shown that such choice of the regularization parameter is optimal one for the solution of our IP since it gives smallest relative error in the reconstruction of the function $c(x)$. We refer to [1, 2, 18], and references therein, for different techniques for the choice of a regularization parameter. The tolerance θ at step 3 of our algorithm of Sect. 5.3 is set to $\theta = 10^{-6}$.

In our numerical simulations we have considered an additive noise σ introduced to the simulated boundary data $\tilde{u}$ in (4) as

$$u_\sigma \left(x^i, t^j\right) = \tilde{u} \left(x^i, t^j\right) \left[1 + \frac{\sigma}{100}\right]. \tag{50}$$

Here, $x^i \in \partial\Omega$ is a mesh point at the boundary $\partial\Omega$, $t^j \in (0, T)$ is a mesh point in the time mesh J_τ, and σ is the noise level in percents.

We use a post-processing procedure to get images of Figs. 4, 7, 10, 11 and 12. This procedure is as follows: assume, that the functions $c^m(x)$ are our reconstructions obtained by the algorithm of Sect. 5.3 where m is the number of iterations in the conjugate gradient algorithm when we have stopped to compute $c(x)$. Then to get our final images, we set

$$\widetilde{c}^m(x) = \begin{cases} c^m(x), & \text{if } c^m(x) > P \max\limits_{\Omega_{FEM}} c^m(x), \\ 1, & \text{otherwise.} \end{cases} \tag{51}$$

The values of the parameter $P \in (0, 1)$ depends on the concrete reconstruction of the function $c(x)$ and plays the roll of a cut-off parameter for the function $c(x)$. If we choose $P \approx 1$ then we will cut almost all reconstruction of the function $c(x)$. Thus,

Table 1 Computational results of the reconstructions in cases (i)–(iv) together with computational errors in achieved contrast in percents. Here, $\overline{N}$ is the final iteration number m in the conjugate gradient method of Sect. 5.3

$\sigma = 3\%$				$\sigma = 10\%$			
Case	$\max_{\Omega_{FEM}} c_{\overline{N}}$	Error, %	$\overline{N}$	Case	$\max_{\Omega_{FEM}} c_{\overline{N}}$	Error, %	$\overline{N}$
(i)	2.21	44.75	7	(i)	3.13	21.75	9
(ii)	2.07	48.25	7	(ii)	3.06	23.5	9
(iii)	5.91	1.5	12	(iii)	4.84	19.3	16
(iv)	5.09	15.2	15	(iv)	5.87	2.2	18

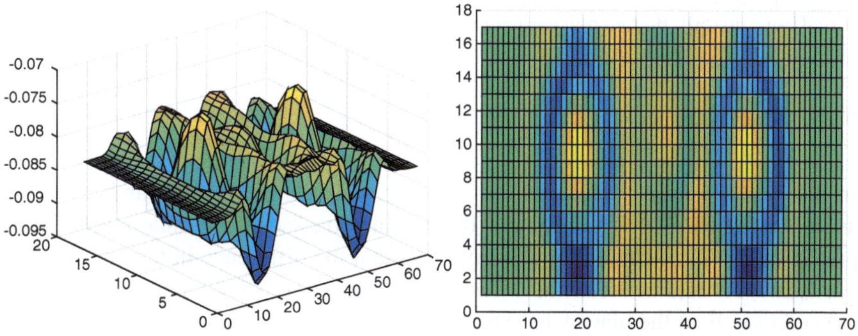

(a) prospect view (b) x_1x_2 view

Fig. 3 Test case (i). Behavior of the noisy backscattered data at time $t = 1.8$ with $\sigma = 3\%$ in (50)

values of P should be chosen numerically. For tests (i), (ii) we have used $P = 0.7$ and for case studies (iii)–(iv) we choose $P - 0.5$.

6.1 Test Case (i)

In this example we performed computations with two noise levels in data: $\sigma = 3\%$ and $\sigma = 10\%$. Figure 3 presents typical behavior of noisy backscattered data in this case. The results of reconstruction for both noise levels are presented in Fig. 4. We observe that the location of all inclusions in x_1x_2 direction is imaged very well. However, the location in the x_3 direction should still be improved.

It follows from Fig. 4 and Table 1 that the imaged contrast in the function $c(x)$ is $2.21 : 1 = \max_{\Omega_{FEM}} c_7 : 1$, where $n := \overline{N} = 7$ is our final iteration number in the conjugate gradient method. Similar observation is valid from Fig. 4 and Table 1 for noise level 10% where imaged contrast in the function $c(x)$ is $3.13 : 1 = \max_{\Omega_{FEM}} c_9 : 1$, where $n := \overline{N} = 9$ is our final iteration number.

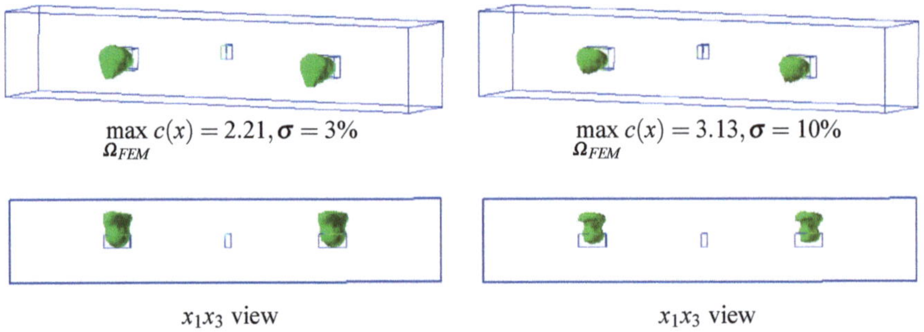

Fig. 4 Test case (i). Computed images of reconstructed $\tilde{c}$ for $\omega = 40$ in (48) and for a different noise level σ in (50). Bottom row present the respective x_1x_3 views

6.2 *Test Case (ii)*

In this test, we have considered the same noise levels σ: $\sigma = 3\%$ and $\sigma = 10\%$, as in the test case (i). The behavior of the noisy backscattered data in this case is presented in Fig. 5. Using Fig. 6 we observe that the difference in the amplitude of backscattered data between the cases (i) and (ii) is very small and, as expected, is located exactly at the place where the middle smallest inclusion of Fig. 2 is moved, This is because the difference in two geometries of Fig. 2 is only in the location of the small middle inclusion: in Fig. 2b this inclusion is moved more close to the backscattered boundary $\partial_1 \Omega$ than in the Fig. 2a.

The results of the reconstruction for both noise levels are presented in Fig. 7. It follows from Fig. 7 and Table 1 that the imaged contrast in the function $c(x)$ is $2.07 : 1 = \max_{\Omega_{FEM}} c_7 : 1$, where $n := \overline{N} = 7$ is our final iteration number in the conjugate gradient method when the noise level is 3 %. Similar observation is valid from Fig. 7 and table 1 for noise level 10 % where imaged contrast in the function $c(x)$ is $3.06 : 1 = \max_{\Omega_{FEM}} c_9 : 1$, where $n := \overline{N} = 9$ is our final iteration number. Again, as in the case (i) we observe that the location of all inclusions in x_1x_2 direction is imaged very well. However, location in x_3 direction should still be improved. We also observe that the smallest inclusion of Fig. 2b is reconstructed better than in the case (i) since it is located closer to the observation boundary $\partial_1 S$. Similar to [5, 8], in our future research, we plan to apply an adaptive finite element method which hopefully will improve the shapes and sizes of all inclusions considered in tests (i) and (ii).

6.3 *Test Case (iii)*

In this numerical test, we reconstruct the conductivity function $c(x)$ which is defined as follows

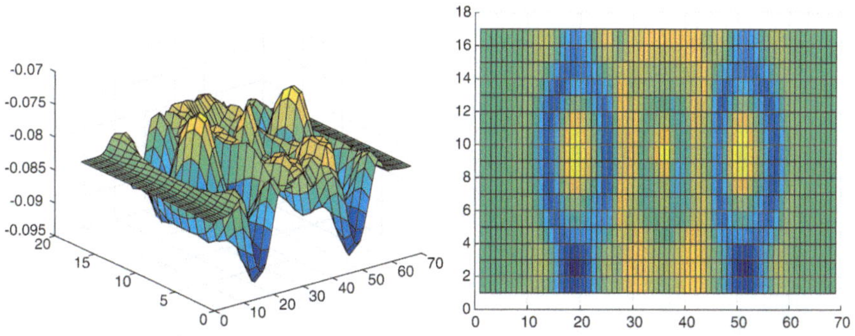

(a) prospect view (b) x_1x_2 view

Fig. 5 Test case (ii). Behavior of the noisy backscattered data at time $t = 1.8$ with $\sigma = 3\%$ in (50)

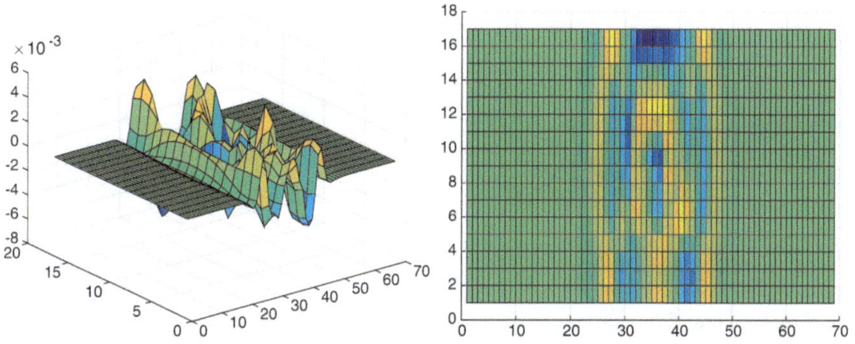

(a) prospect view (b) x_1x_2 view

Fig. 6 The difference of the noisy backscattered data at time $t = 1.8$ in case studies (i) and (ii) when noise level is $\sigma = 3\%$ in (50)

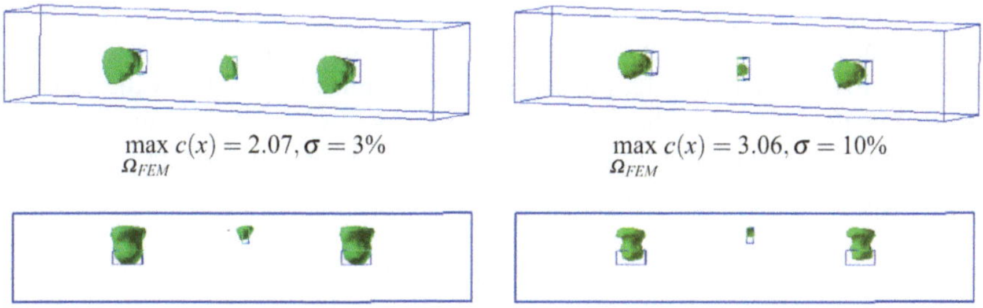

$\max_{\Omega_{FEM}} c(x) = 2.07, \sigma = 3\%$ $\max_{\Omega_{FEM}} c(x) = 3.06, \sigma = 10\%$

x_1x_3 view x_1x_3 view

Fig. 7 Test case (ii). Computed images of reconstructed $\tilde{c}$ for $\omega = 40$ in (48) and for different noise level σ in (50). Bottom row present the x_1x_3 views

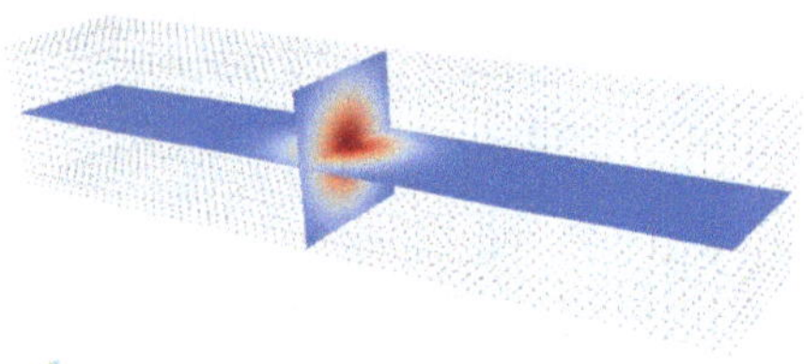

(a) Test (iii): horizontal and vertical slices

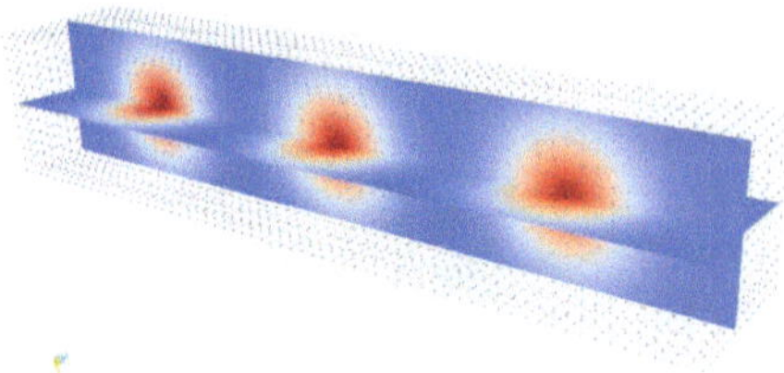

(b) Test (iv): horizontal and vertical slices

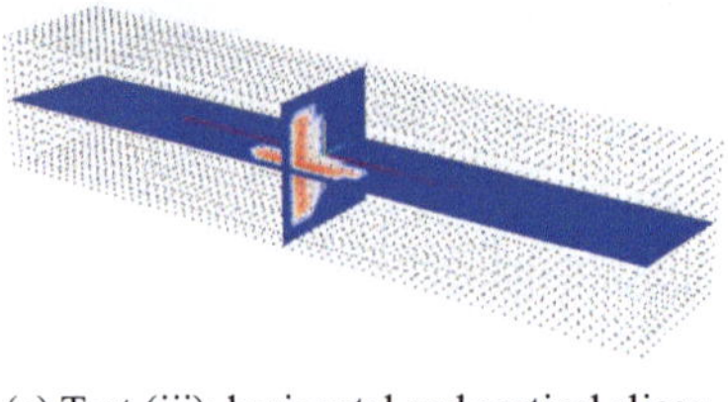

(c) Test (iii): horizontal and vertical slices

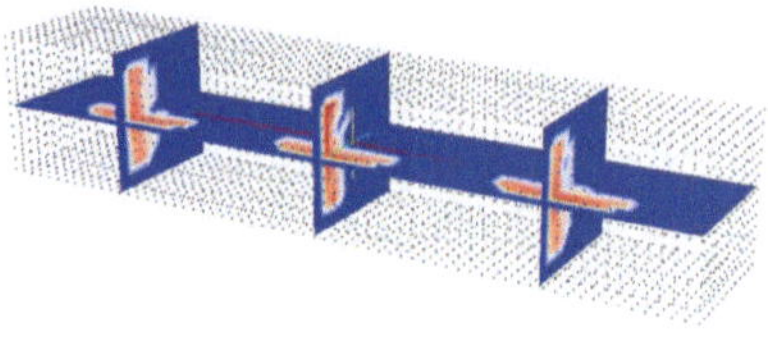

(d) Test (iv): horizontal and vertical slices

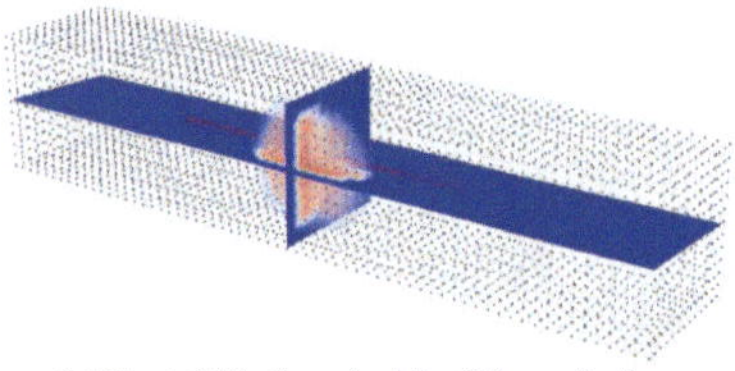

(e)Test (iii): threshold of the solution

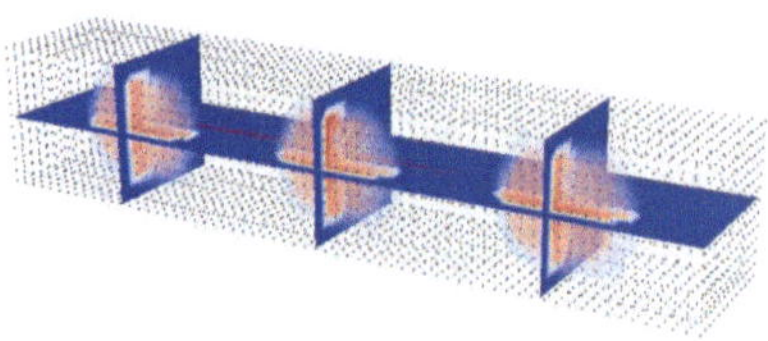

(f) Test (iv): threshold of the solution

Fig. 8 **a**, **b**. Slices of the exact Gaussian functions given by (52) and (54), respectively. **c**, **d** Slices and **e**, **f** thresholds of the reconstructions. Here, computations were done for the noise $\sigma = 10\%$ and $\omega = 40$

$$c(x) = 1.0 + 5.0 \cdot \exp^{-(x_1{}^2/0.2 + x_2{}^2/0.2 + x_3{}^2/0.2)}, \tag{52}$$

see Fig. 8a. In this test, we have used noisy boundary data u_σ with $\sigma = 3\%$ and $\sigma = 10\%$ in (50). Note that a priori we have not assumed that we know the structure of this function, ruther we have assumed that we know the lower bound $c(x) \geq 1$ and that the reconstructed values of the conductivity belongs to the set of admissible parameters which is now defined as

$$M_c \in \{c \in C(\overline{\Omega}) | 1 \leq c(x) \leq 10\}. \tag{53}$$

Figures 8c, e and 10 display results of the reconstruction of function given by (52) with $\sigma = 10\%$ in (50). Quite similar results are obtained for $\sigma = 3\%$ in (50), see Fig. 9. We observe that the location of the maximal value of the function (50) is imaged very well. It follows from Fig. 9 and Table 1 that the imaged contrast in this function is $5.91 : 1 = \max_{\Omega_{FEM}} c_{12} : 1$, where $n := \overline{N} = 12$ is our final iteration number in the conjugate gradient method. Similar observation is valid from Fig. 10 and Table 1 where the imaged contrast is $4.84 : 1 = \max_{\Omega_{FEM}} c_{16} : 1, n := \overline{N} = 16$.

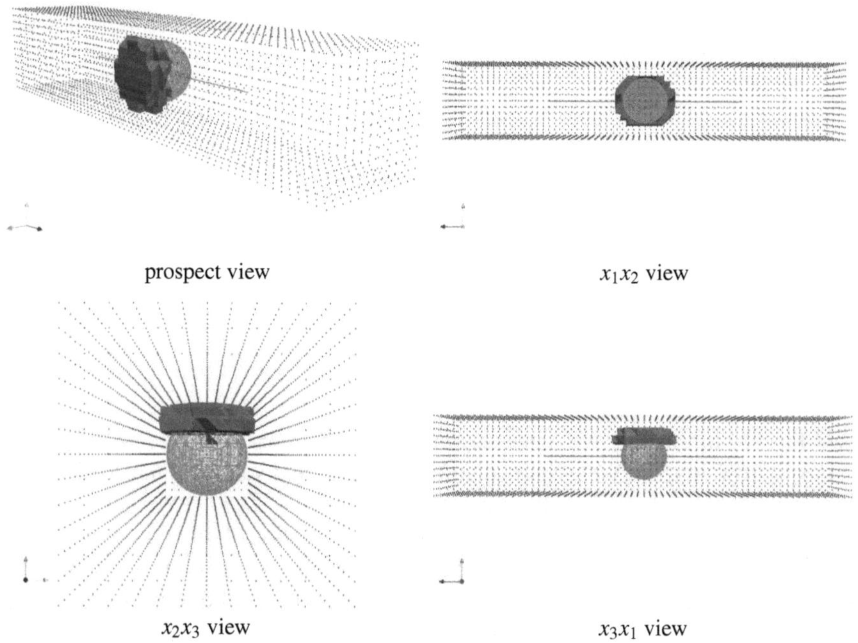

Fig. 9 Test case (iii). We present reconstruction of $\tilde{c}$ with $\max_{\Omega_{FEM}} c(x) = 5.91$ for $\omega = 40$ in (48) with noise level $\sigma = 3\%$. The spherical wireframe of the isosurface with exact value of the function (52), which corresponds to the value of the reconstructed $\tilde{c}$, is outlined by a thin line

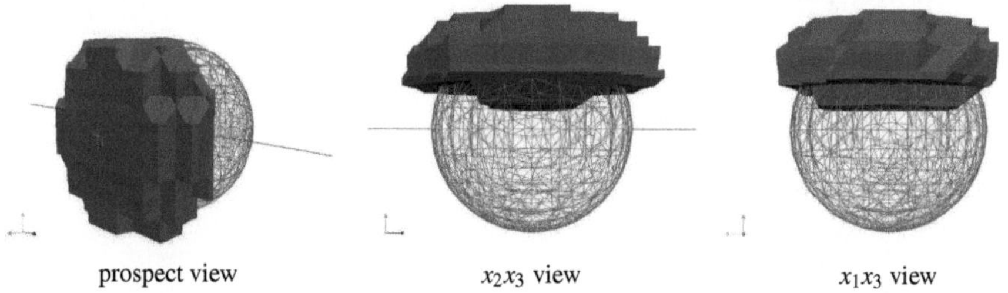

Fig. 10 Test case (iii). Computed images of reconstructed $\max_{\Omega_{FEM}} c(x) = 4.84$ for $\omega = 40$ in (48) and noise level $\sigma = 10\%$. The spherical wireframe of the isosurface with exact value of the function (52), corresponding to the value of the reconstructed $\tilde{c}$, is outline by a thin line

However, from these figures we also observe that because of the data post-processing procedure (51) the values of the background $1.0 + 5.0 \cdot \exp^{-(x^2/0.2+y^2/0.2+z^2/0.2)}$ in (52) are not reconstructed but are smoothed out. Thus, we are able to reconstruct only maximal values of the function (52). Comparison of Figs. 8c, e, 9, 10 with Fig. 8a reveals that it is desirable to improve shape of the function (52) in x_3 direction. Again, similar to [5, 8] we hope that an adaptive finite element method can refine the obtained images of Fig. 10 in order to get better shapes and sizes of the function (52) in all directions.

6.4 Test Case (iv)

In our last numerical test we reconstruct the conductivity function $c(x)$ given by three sharp Gaussians such that

$$\begin{aligned} c(x) &= 1.0 + 5.0 \cdot \exp^{-((x_1+2)^2/0.2+x_2{}^2/0.2+x_3{}^2/0.2)} \\ &+5.0 \cdot \exp^{-(x_1{}^2/0.2+x_2{}^2/0.2+x_3{}^2/0.2)} + 5.0 \cdot \exp^{-((x_1-2)^2/0.2+x_2{}^2/0.2+x_3{}^2/0.2)}, \end{aligned} \tag{54}$$

see Fig. 8b. In this test we again used the noisy boundary data u_σ with $\sigma = 3\%$ and $\sigma = 10\%$ in (50). We assume that the reconstructed values of the conductivity belongs to the set of admissible parameters (53).

Figures 8d, f, 11 and 12 show results of the reconstruction of function given by (54) for $\sigma = 3\%$ and $\sigma = 10\%$ in (50), respectively. We observe that the location of the maximal value of the function (54) is imaged very well. It follows from Fig. 11 and

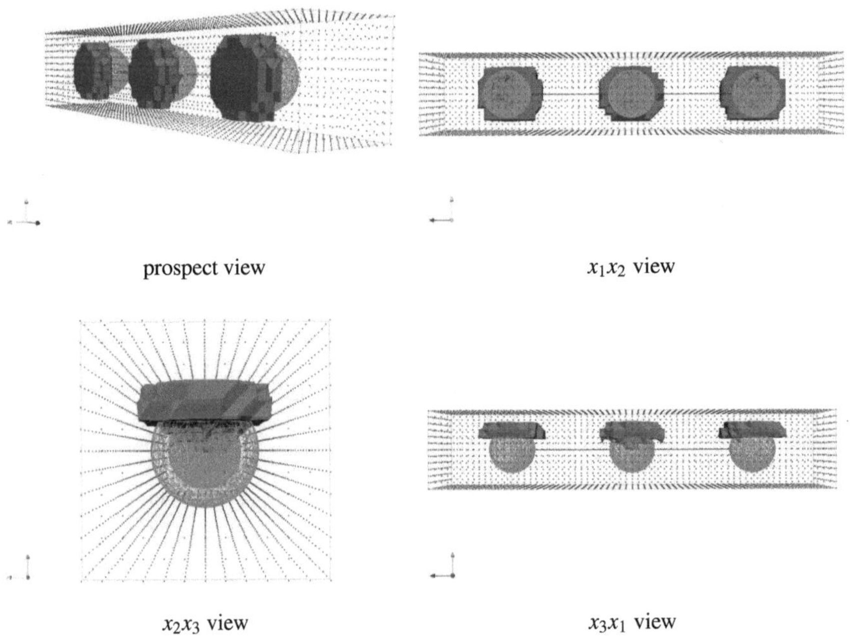

Fig. 11 Test case (iv). We present reconstruction of $\tilde{c}$ when $\max_{\Omega_{FEM}} c(x) = 5.09$ for $\omega = 40$ in (48) and noise level $\sigma = 3\%$. The spherical wireframe of the isosurface with exact value of the function (54), which corresponds to the value of the reconstructed $\tilde{c}$, is outlined by a thin line

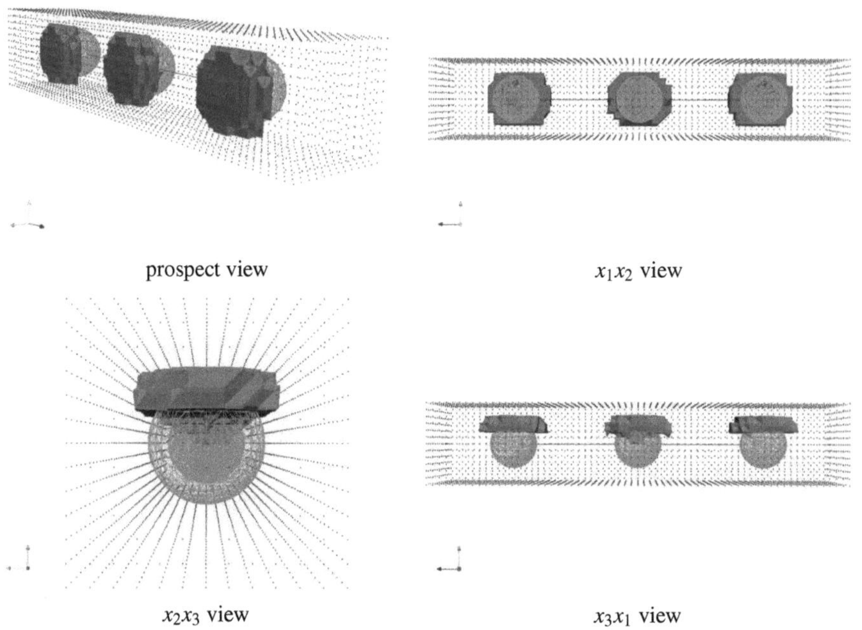

Fig. 12 Test case (iv). We present reconstruction of $\tilde{c}$ when $\max_{\Omega_{FEM}} c(x) = 5.87$ for $\omega = 40$ in (48) and noise level $\sigma = 10\%$. The spherical wireframe of the isosurface with exact value of the function (54), corresponding to the value of the reconstructed $\tilde{c}$, is outlined by a thin line

Table 1 that when the noise level is $\sigma = 3\%$ then the imaged contrast in this function is $5.09 : 1 = \max_{\Omega_{FEM}} c_{15} : 1$, where $n := \overline{N} = 15$ is our final iteration number in the conjugate gradient method. When the noise level is $\sigma = 10\%$ then the imaged contrast is $5.87 : 1 = \max_{\Omega_{FEM}} c_{18} : 1, n := \overline{N} = 18$.

However, as in the case (iii), the values of the background in (54) are not reconstructed but are smoothed out and we are able to reconstruct only maximal values of the three Gaussians in the function (54). Comparing Figs. 8d, f, 11, 12 with Fig. 8b we see that it is desirable to improve shapes of the function (54) in x_3 direction.

7 Discussion and Conclusion

In this work, we have presented a computational study of the reconstruction of the conductivity function $c(x)$ in a hyperbolic problem (1) using Lagrangian approach and a hybrid finite element/difference method of [3]. As theoretical result, we have presented estimate of the norms between computed and regularized solutions of the Tikhonov functional via the L_2 norm of the Fréchet derivative of this functional or via the corresponding Lagrangian.

In our numerical tests, we have obtained stable reconstruction of the conductivity function $c(x)$ in $x_1 x_2$-directions for frequency $\omega = 40$ in the initialization of a plane wave (48) and for noise levels $\sigma = 3\%, 10\%$ in (50). However, size and shape on x_3 direction should still be improved in all test cases. Similar to [5, 8] we plan to apply an adaptive finite element method in order to get better shapes and sizes of the conductivity function $c(x)$ in x_3 direction.

Using results of Table 1 we can conclude that the computational errors in the achieved maximal contrast are less in the case of reconstruction of smooth functions than in the reconstruction of small inclusions. This can be explained by involving of discontinuities in the reconstruction of small inclusions, as well as by having special geometry in these small inclusions: all of them have different sizes and locations inside Ω_{FEM}, and thus, achieving the exact contrast becomes more difficult task in this case. The important observation is that when the scatters are of different size, especially when the smallest scatterer is located between larger ones, as in case studies (i) and (ii), we note that the smaller scatterer is better reconstructed when it is located near the observation boundary of the computational domain Ω_{FEM}.

Acknowledgements The research of L. B. is partially supported by the sabbatical programme at the Faculty of Science, University of Gothenburg, Sweden, and the research of K. N. was supported by the Swedish Foundation for Strategic Research. The computations were performed on resources at Chalmers Centre for Computational Science and Engineering (C3SE) provided by the Swedish National Infrastructure for Computing (SNIC).

References

1. A.B. Bakushinsky, M.Y. Kokurin, *Iterative Methods for Approximate Solution of Inverse Problems* (Springer, 2004)
2. A. Bakushinsky, M.Y. Kokurin, A. Smirnova, *Iterative Methods for Ill-posed Problems*, vol. 54, Inverse and Ill-Posed Problems Series (De Gruyter, 2011)
3. L. Beilina, Domain Decomposition finite element/finite difference method for the conductivity reconstruction in a hyperbolic equation. Commun Nonlinear Sci Numer Simul. Elsevier (2016). https://doi.org/10.1016/j.cnsns.2016.01.016
4. L. Beilina, K. Samuelsson, K. Åhlander, Efficiency of a hybrid method for the wave equation, in *Proceedings of the International Conference on Finite Element Methods: Three dimensional problems*. GAKUTO International Series, Mathematical Sciences and Applications, **15** (2001)
5. L. Beilina, C. Johnson, A posteriori error estimation in computational inverse scattering. Math Models Appl Sci **1**, 23–35 (2005)
6. L. Beilina, M.V. Klibanov, *Approximate Global Convergence and Adaptivity for Coefficient Inverse Problems* (Springer, New-York, 2012)
7. L. Beilina, M.V. Klibanov, MYu. Kokurin, Adaptivity with relaxation for ill-posed problems and global convergence for a coefficient inverse problem. J Math Sci **167**, 279–325 (2010)
8. L. Beilina, N.T. Thành, M.V. Klibanov, J. Bondestam-Malmberg, Reconstruction of shapes and refractive indices from blind backscattering experimental data using the adaptivity. Inverse Probl. **30**, 105007 (2014)
9. L. Beilina, M. Cristofol, K. Niinimäki, Optimization approach for the simultaneous reconstruction of the dielectric permittivity and magnetic permeability functions from limited observations. Inverse Probl Imaging **9**(1), 1–25 (2015)

10. C. Bellis, M. Bonnet, B.B. Guzina, Apposition of the topological sensitivity and linear sampling approaches to inverse scattering. Wave Motion **50**, 891–908 (2013)
11. S.C. Brenner, L.R. Scott, *The Mathematical Theory of Finite Element Methods* (Springer, Berlin, 1994)
12. H.D. Bui, A. Constantinescu, H. Maigre, Numerical identification of planar cracks in elastodynamics using the instantaneous reciprocity gap. Inverse Probl **20**, 993–1001 (2004)
13. H.D. Bui, A. Constantinescu, H. Maigre, An exact inversion formula for determining a planar fault from boundary measurements. J Inverse Ill Posed Probl **13**, 553–565 (2005)
14. Y.T. Chow, J. Zou, A numerical method for reconstructing the coefficient in a wave equation. Numer Methods Partial Differ Equ **31**, 289–307 (2015)
15. G.C. Cohen, *Higher Order Numerical Methods for Transient Wave Equations* (Springer, 2002)
16. R. Courant, K. Friedrichs, H. Lewy, On the partial differential equations of mathematical physics. IBM J Res Dev **11**(2), 215–234 (1967)
17. M. Cristofol, S. Li, E. Soccorsi, Determining the waveguide conductivity in a hyperbolic equation from a single measurement on the lateral boundary. Math Control Relat Fields **6**(3), 407–427 (2016)
18. H.W. Engl, M. Hanke, A. Neubauer, *Regularization of Inverse Problems* (Kluwer Academic Publishers, Boston, 2000)
19. B. Engquist, A. Majda, Absorbing boundary conditions for the numerical simulation of waves. Math. Comp. **31**, 629–651 (1977)
20. S.N. Fata, B.B. Guzina, A linear sampling method for near-field inverse problems in elastodynamics. Inverse Probl **20**, 713–736 (2004)
21. M.V. Klibanov, A.B. Bakushinsky, L. Beilina, Why a minimizer of the Tikhonov functional is closer to the exact solution than the first guess. J Inverse Ill Posed Probl **19**(1), 83–105 (2011)
22. O.A. Ladyzhenskaya, *Boundary Value Problems of Mathematical Physics* (Springer, Berlin, 1985)
23. PETSc, Portable, Extensible Toolkit for Scientific Computation. http://www.mcs.anl.gov/petsc/
24. O. Pironneau, *Optimal Shape Design for Elliptic Systems* (Springer, Berlin, 1984)
25. A.N. Tikhonov, A.V. Goncharsky, V.V. Stepanov, A.G. Yagola, *Numerical Methods for the Solution of Ill-Posed Problems* (Kluwer, London, 1995)
26. WavES, the software package. http://www.waves24.com

On Finite Element Method for Magnetic Resonance Imaging

L. Beilina, G. Guillot and K. Niinimäki

Abstract In this work we consider the problem of magnetic resonance imaging (MRI). We propose and formulate a finite element method to handle this problem and present an adaptive algorithm for local mesh refinements. Reconstructions from experimental MR data are shown and used to compare different interpolation techniques.

Keywords Magnetic resonance imaging · Finite element method · Tikhonov functional · Adaptivity · Interpolation · Experimental data

1 Introduction

In this work we study the MRI problem which is described by a Fredholm integral equation of the first kind. For the solution of this problem we formulate a finite element method (FEM) and an adaptive algorithm for local mesh refinements.

An adaptive finite element method for a Fredholm integral equation of the second kind was considered in [1]. Solution of a Fredholm integral equation of the first kind is a more difficult problem since it is an ill-posed problem. To approach it minimization of the Tikhonov functional should be used. In this setting, an adaptive FEM for a Fredholm integral equation of the first kind was proposed in [8], where the authors

L. Beilina (✉)
Department of Mathematical Sciences, Chalmers University of Technology and University of Gothenburg, 41296 Gothenburg, Sweden
e-mail: larisa.beilina@chalmers.se

G. Guillot
IR4M UMR8081, CNRS, Université Paris-Sud, Université Paris-Saclay, bâtiment 220, 4 place du Général Leclerc, 91401 Orsay, France
e-mail: genevieve.guilloti@u-psud.fr

K. Niinimäki
IR4M UMR8081, CNRS, Université Paris-Sud, Université Paris-Saclay, SHFJ, 4 place du Général Leclerc, 91401 Orsay, France
e-mail: kati.niinimaki@u-psud.fr

L. Beilina and Yu. G. Smirnov (eds.), *Nonlinear and Inverse Problems in Electromagnetics*, Springer Proceedings in Mathematics & Statistics 243,
https://doi.org/10.1007/978-3-319-94060-1_9

also derived a posteriori error estimates for the minimization functional and for the regularized solution of this functional. However, similar to the study in [12], the adaptive algorithm of [8] relies on the discretized version of the minimizer using the convolution theorem. The novelty of the present work is that we have formulated the finite element method in the discrete form without using the convolution theorem, and we propose new adaptive algorithm. This algorithm uses only finite element formulation for the minimization of the Tikhonov functional.

In the numerical part of our work we first present reconstructions that were obtained from experimental MR data. The data was acquired at IR4M laboratory in Paris-Sud University, France. We also study the effect of different interpolation techniques, which are related to the proposed adaptive algorithm, by applying them to the reconstructed MR images.

The outline of this paper is as follows. In Sect. 2 we present the statement of the forward and inverse problem in MRI. In Sect. 3 we formulate the finite element method for minimization of the Tikhonov functional. Section 4 describes our adaptive algorithm and interpolation techniques. Finally, Sect. 5 presents numerical results and Sect. 6 gives conclusions.

2 Statement of the Forward and Inverse Problem

Let $\Omega \subset \mathbb{R}^m$, $m = 2, 3$ be a bounded domain such that $\Omega := [0, a] \times [0, b]$ for $m = 2$ and $\Omega := [0, a] \times [0, b] \times [0, c]$ for $m = 3$, with the boundary $\partial\Omega$. Further, let $\Omega_\kappa \subset \mathbb{R}^m$ be another bounded domain of the size $\Omega_\kappa := [0, a_\kappa] \times [0, b_\kappa]$ for $m = 2$ and $\Omega_\kappa := [0, a_\kappa] \times [0, b_\kappa] \times [0, c_\kappa]$ for $m = 3$. We call the domain Ω image-space and the domain Ω_κ k-space. The data is collected in Ω_κ.

Let H denote the Hilbert space of all functions defined in the domain Ω. Next, let function $w(x), x = (x_1, ..., x_m) \in \Omega$ be a k times continuously differentiable in Ω. We denote the partial derivative of the order $|\alpha| \leq k$, of function w by $D^\alpha w$.

We denote by $C^k\left(\overline{\Omega}\right)$ the Banach space of functions $w(x)$ which are continuous in the closure $\overline{\Omega}$ of the domain Ω together with their derivatives $D^\alpha w$, $|\alpha| \leq m$. The norm in this space is defined as

$$\|w\|_{C^k(\Omega)} = \sum_{|\alpha| \leq m} \sup_{x \in \Omega} |D^\alpha w(x)| < \infty.$$

Consider the Sobolev space $H^k(\Omega)$ of all functions with the norm defined as

$$\|w\|^2_{H^k(\Omega)} = \sum_{|\alpha| \leq k} \int_\Omega |D^\alpha w|^2 \, dx < \infty,$$

where $D^\alpha w$ are weak derivatives of the function w.

$H^k(\Omega)$ is a Hilbert space with the inner product defined as

$$(w, v)_{H^k(\Omega)} = \sum_{|\alpha|\leq k} \int_\Omega D^\alpha w \, D^\alpha v \, dx.$$

In $L_2(\Omega)$ the inner product and the norm are defined as

$$(w, v) = \int_\Omega wv \, dx, \ x \in \Omega,$$
$$||w||^2 = (w, w).$$

We consider solution of a two-dimensional Fredholm integral equation of the first kind

$$u(k_x, k_y) = \int_\Omega f(x, y) G(x, y, k_x, k_y) \, \mathrm{d}x\mathrm{d}y, \tag{1}$$

where $f(x, y) \in H(\Omega)$ denotes the unknown image and $u(k_x, k_y) \in L_2(\Omega_\kappa)$ is the experimental MR data and $G(x, y, k_x, k_y) = e^{-2\pi \mathbf{i}(k_x x + k_y y)} \in C^k, k > 0$ denotes the kernel and $\mathbf{i}$ denotes the imaginary unit. The above equation can be extended to the three-dimensional Fredholm integral equation of the first kind

$$u(k_x, k_y, k_z) = \int_\Omega f(x, y, z) G(x, y, z, k_x, k_y, k_z) \, \mathrm{d}x\mathrm{d}y\mathrm{d}z, \tag{2}$$

where $f(x, y, z) \in H(\Omega)$ denotes the unknown object, $u(k_x, k_y, k_z) \in L_2(\Omega_\kappa)$ is the experimental MR data and $G(x, y, z, k_x, k_y, k_z) = e^{-2\pi \mathbf{i}(k_x x + k_y y + k_z z)} \in C^k, k > 0$. In these equations, (k_x, k_y, k_z) denote the k-space trajectories which define how the k-space data in Ω_κ is acquired and (k_x, k_y, k_z) correspond to the coordinates of $u \in \Omega_\kappa$. For more details about description of MRI problem we refer to [6].

The Eqs. (1) and (2) can be written in the operator form

$$Af = u, \tag{3}$$

with a bounded linear operator $A : H(\Omega) \to L_2(\Omega_\kappa)$ defined for $m = 2$ as

$$Af := \int_\Omega f(x, y) e^{-2\pi \mathbf{i}(k_x x + k_y y)} \, \mathrm{d}x\mathrm{d}y, \tag{4}$$

and for $m = 3$ as

$$Af := \int_\Omega f(x, y, z) e^{-2\pi \mathbf{i}(k_x x + k_y y + k_z z)} \, \mathrm{d}x\mathrm{d}y\mathrm{d}z. \tag{5}$$

Below we formulate the image reconstruction problems in two and three dimensions. The following inverse problems are either mildly or strongly ill-posed depending on the data. Ill-posedness means high sensitivity to inaccuracies in modeling and to errors in measurement. Overcoming this sensitivity requires regularization, see details in [2, 3, 10–12].

Inverse Problem in 2D (IP2D)

Find $f(x, y)$ in (1) when the measured MR data $u(k_x, k_y) \in \Omega_\kappa$, the k-space coordinates (k_x, k_y) and the kernel $G(x, y, k_x, k_y)$ are known.

Inverse Problem in 3D (IP2D)

Find $f(x, y, y)$ in (2) when the measured MR data $u(k_x, k_y, k_z) \in \Omega_\kappa$, the k-space coordinates (k_x, k_y, k_z) and the kernel $G(x, y, z, k_x, k_y, k_z)$ are known.

Let W_1,W_2, Q be three Hilbert spaces such that $Q \subseteq W_1$ as a set, and the norm in Q is stronger than the norm in W_1, $\overline{Q} = W_1$. To find solution of (IP2D) or (IP3D) we construct the Tikhonov regularization functional

$$M_\gamma(f) = \frac{1}{2} \|Af - u\|_{W_2}^2 + \frac{\gamma}{2} \|f\|_Q^2, \tag{6}$$

$$M_\gamma(f) : Q \to \mathbb{R}, \;\; u \in W_2, \, f \in Q,$$

and we want to find the minimum of it

$$\inf_{f \in Q} \{M_\gamma(f)\} = \inf_{f \in Q} \{\frac{1}{2} \|Af - u\|_{W_2}^2 + \frac{\gamma}{2} \|f\|_Q^2\}$$

such that for all $b \in Q$

$$M'_\gamma(f)(b) = 0, \tag{7}$$

where $M'_\gamma(f)$ denotes the Fréchet derivative of the functional (6) and $M'_\gamma(f)(b)$ means that M'_γ acts on b. Here, $\gamma = \gamma(\delta) > 0$ is a regularization parameter depending on the noise δ in data such that

$$\|u - u^*\|_{L_2(\Omega_\kappa)} \le \delta,$$

where u^* denote perfect noiseless data corresponding to the exact solution f^* of (3) such that

$$Af^* = u^*. \tag{8}$$

Below we present the lemma from [3] when the operator $A : L_2(\Omega) \to L_2(\Omega_\kappa)$.

Lemma 1 *Let $A : L_2(\Omega) \to L_2(\Omega_\kappa)$ be a bounded linear operator. Then the Fréchet derivative of the functional (6) is*

$$M'_\gamma(f)(b) = (A^*Af - A^*u, b) + \gamma(f, b), \; \forall b \in Q. \tag{9}$$

3 The Finite Element Method for Minimization of the Tikhonov Functional

To formulate the finite element method for (7) we discretize the domain $\Omega \subset \mathbb{R}^m, m = 2, 3$ by an unstructured mesh K_h. This mesh consists of non-overlapping elements K which are tetrahedral or hexahedral elements in $\mathbb{R}^3$ and triangles or quadrilateral in $\mathbb{R}^2$ such that $K_h = K_1, ..., K_s$, where s is the number of elements in Ω, and

$$\Omega = \cup_{K \in K_h} K = K_1 \cup K_2 ... \cup K_s.$$

We determine the mesh function $h = h(x)$ which is a piecewise-constant function such that

$$h(x) = h_K \quad \forall K \in K_h,$$

where h_K is the diameter of K which we define as the longest side of K. As usual, see details in [7], we let r' be the radius of the maximal circle/sphere contained in the element K. We make the following shape regularity assumption for every element $K \in K_h$

$$c_1 \le h' \le r' c_2; \quad c_1, c_2 = const. > 0. \tag{10}$$

For the data u, we use discretization $u = \big(u_1(k_{x_1}, k_{y_1}), \ldots, u_M(k_{x_M}, k_{y_M})\big)$ with coordinates (k_x, k_y) in Ω_κ, when $m = 2$ and $u = \big(u_1(k_{x_1}, k_{y_1}, k_{z_1}), \ldots, u_M(k_{x_M}, k_{y_M}, k_{z_M})\big)$ with coordinates (k_x, k_y, k_z) in Ω_κ, when $m = 3$. In the following we denote by $u_i, i = 1, \ldots, M$ the known discretized measured data in the k-space Ω_κ.

We define the finite element space $V_h \subset V$ as

$$V_h = \big\{v \in V : v \in C(\Omega),\ v|_K \in P_1(K)\ \forall K \in K_h\big\}, \tag{11}$$

where $P_1(K)$ denotes the set of piecewise-linear functions on K with

$$V = \big\{v : v \in L_2(\Omega)\big\}.$$

Let us consider the Eq. (3) in the form

$$A^* A f = A^* u. \tag{12}$$

We approximate function f by $f_h \in V_h$, such that

$$f_h = \sum_{i=1}^{N} f_i \varphi_i, \tag{13}$$

where f_i denote the unknown discrete function-values at the mesh point $x_i \in K_h$. Here, $\{\varphi_i\}_{i=1}^N$ are the standard continuous piecewise linear functions in space such that

$$(\varphi_i, \varphi_j) = \delta_{ij}.$$

The finite element method for the Eq. (12) is: find $f_h \in V_h$ such that for all $v \in V_h$

$$(A^* A f_h, v) = (A^* u, v). \tag{14}$$

Substituting discrete representations for $u = (u_1, \ldots, u_M)$ and for f as in (13), into (14) with $v = \varphi_j$ we obtain the system of discrete linear equations:

$$\sum_{i,j=1}^{N} (A\varphi_i, A\varphi_j) f_i = \sum_{i,j=1}^{N} (u_i, A\varphi_j). \tag{15}$$

Further, for convenience, we will use $M = N$ for u.

The finite element method for (7) reads: find $f_h \in V_h$ such that for all $v \in V_h$

$$M'_\gamma(f_h)(v) = (A^* A f_h - A^* u, v) + \gamma (f_h, v) = 0. \tag{16}$$

Substituting (13) and (15) into the expression above we get the discrete system of equations

$$\sum_{i,j=1}^{N} (A\varphi_i, A\varphi_j) f_i - \sum_{i,j=1}^{N} (u_i, A\varphi_j) + \gamma \sum_{i,j=1}^{N} (\varphi_i, \varphi_j) f_i = 0, \tag{17}$$

which is equivalent to the system of linear equations

$$\sum_{i,j=1}^{N} (A\varphi_i, A\varphi_j) f_i + \gamma \sum_{i,j=1}^{N} (\varphi_i, \varphi_j) f_i = \sum_{i,j=1}^{N} (u_i, A\varphi_j). \tag{18}$$

For the fully discrete scheme on the finite element mesh K_h the system (18) can be written as

$$\sum_{K \in K_h} \sum_{i,j=1}^{N} \int_K A\varphi_i A\varphi_j dK \ f_i + \gamma \sum_{K \in K_h} \sum_{i,j=1}^{N} \int_K \varphi_i \varphi_j dK \ f_i = \sum_{K \in K_h} \sum_{i,j=1}^{N} \int_K u_i A\varphi_j dK \ , \tag{19}$$

or as

$$(C + \gamma M)\mathbf{f} = B. \tag{20}$$

Here, C and M are the block matrices, $\mathbf{f}$ denote the nodal values of f_h and B is the assembled vector of the right hand side.

For computation of the entries of C, B and M we will define the mapping F_K for the reference element $\hat{K}$ such that $F_K(\hat{K}) = K$. We also define by $\hat{\varphi}$, the local piecewise linear basis function on the reference element $\hat{K}$ such that $\nabla\varphi \circ F_K = DF_K^{*-1}\nabla\hat{\varphi}$, where DF_K^{*-1} denotes the inverse of the transposed Jacobian matrix of the transformation F_K and $DF_K^* = (DF_K)^T$. Then we can write for every element K in the mesh K_h formulas for the entries in system (20) in the terms of global test functions $\varphi_{i,j}, i, j = 1, \ldots, N$

$$
\begin{aligned}
C_{i,j}^K &= (A\varphi_i \circ F_K,\ A\varphi_j \circ F_K)_K, \\
B_j^K &= (u_i,\ A\varphi_j \circ F_K)_K, \\
M_{i,j}^K &= (\ \varphi_i \circ F_K,\ \varphi_j \circ F_K)_K.
\end{aligned} \tag{21}
$$

4 The Adaptive Algorithm

In this section we present the adaptive algorithm which is based on the finite element solution of system (20) and on the first mesh refinement recommendation of [8]. According to the Theorem 5.1 of [8], we propose to refine the finite element mesh $K_l := (K_h)_l, l = 0, 1, ...M$ in the neighborhood of those points in Ω where the computed function f_h attains its maximal values. Here, l denotes the number of mesh refinements.

In order to obtain the function u_h on the new refined mesh K_l we use interpolation of u_h from the mesh K_{l-1} to the new refined mesh K_l. In [8] a smoothing interpolation procedure was used based on the averaging of the function u_h at every new element of the new mesh K_l over the vertices of the parent triangle of the mesh K_{l-1}.

In our work, we propose to use a barycentric coordinates for interpolation of the function f_h or any other known discrete function, from the mesh K_{l-1} to the new refined mesh K_l. The interpolation is done by

$$
f(r) = \sum_{i=1}^{m+1} \lambda_i f(n_i), \tag{22}
$$

where $m = 2, 3$, $f(n_i)$ are known values of the discrete function f_h at the nodes of the parent element K in the mesh K_{l-1}, and λ_i are barycentric coordinates of the point r which should be inside or at the boundary of the element K. We refer to [9] for computation of the barycentric coordinates.

Adaptive algorithm

- Step 0.
  Generate the initial mesh K_0 in Ω. Compute the sequence of reconstructions $f_l, l \geq 0$ on the refined meshes, as:

- Step 1.
  Compute the finite element solution f_l on K_l by solving the system of linear equations (20).
- Step 2.
  Refine locally the mesh T_l at all points where

$$|f_l| \geq \beta \max_{\Omega} |f_l|, \tag{23}$$

  where $\beta \in (0, 1)$ should be chosen computationally, see, for example, [4] for details.
- Step 3.
  Construct a new mesh K_{l+1} in Ω. Interpolate u_l, f_l to u_{l+1}, f_{l+1}, correspondingly, using barycentric interpolation (22).
- Step 4.
  Compute norms $\|f_{l+1} - f_l\|_{L_2(K_{l+1})}$. Stop mesh refinements when $\|f_{l+1} - f_l\|_{L_2(K_{l+1})} < \epsilon$. Otherwise go to step 1. Here, ϵ is tolerance chosen by the user.

Depending on the application the region for local mesh refinements can be selected differently by the user.

5 Numerical Experiment

In Sect. 5.1 we present the reconstruction of a phantom from experimentally measured MR data. Section 5.2 shows the performance of applying different interpolation techniques on the reconstructed image of Sect. 5.1.

5.1 MR Data Acquisition

For the purpose of this work, a cylindrical phantom (10 mm diameter) was used that consisted of eight cylinders (2 mm diameter) of hardened polymeric bone cement (Osteopal®) in a 0.4 mMol/L $MnCl_2$ water solution. A laboratory-assembled 4.7 T scanner with a 30 cm diameter bore and a gradient system with 12 cm internal diameter and peak gradient amplitude of 400 mT/m were used. The scanner was controlled by a Tecmag sequencer (Apollo, Tecmag, Houston, Texas, USA). The radio-frequency coil was a 20 mm circular loop through which the cylindrical sample was inserted, to put the region of interest in the same plane as the loop. For MR acquisition, a standard 3D Gradient Echo sequence was used with the following parameters: echo time (TE) 2.2 ms, repetition time (TR) 20 ms, flip angle 15°, band width 50 kHz, field-of-view 33 mm × 16.5 mm × 16.5 mm, acquisition matrix 150 × 75 × 75. For further information on MRI acquisition sequences, see [6], for example.

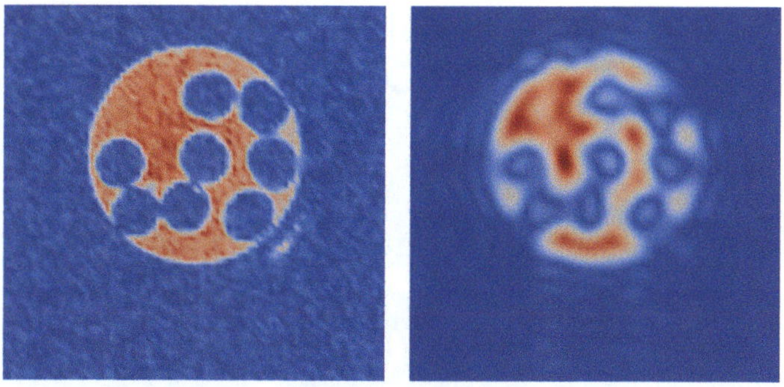

Fig. 1 Original images obtained from MR experiment on initial mesh K_0. These images were used for the evaluation of different interpolation techniques. On the left: image corresponding to the dataset 1. On the right: image corresponding to the dataset 2

Images which were used to evaluate the interpolation methods were reconstructed by discrete inverse Fourier transform (Matlab®'s FFT routines). We considered two sets of image data: dataset 1 and dataset 2. Dataset 1 was obtained as follows: all acquired 3D data in the k-space Ω_κ was used and a 3D image volume was computed via inverse Fourier transform. Then a 2D image slice of size 75×75 was selected which shows the circular cross sections of all of the eight bone cement cylinders. For the dataset 2 we took only the middle part of the k-space and computed a 3D image volume by inverse Fourier transform. Then we selected the same 2D cross section as in dataset 1. The use of only the central part of the k-space results in a blurry image. The images of datasets 1 and 2 are shown in Fig. 1.

5.2 Results Using Different Interpolation Techniques

The goal of this section is to show the performance of the different interpolation techniques described in Sect. 4. These techniques are used in the adaptive algorithm for the interpolation of the reconstructed function f_l from the coarser mesh K_l to the new refined mesh K_{l+1}. A correct interpolation will result in a correct reconstruction of the final image f_M on the finally refined mesh K_M in the algorithm of Sect. 4.

Following our algorithm we generated an initial mesh K_0 for the images of datasets 1 and 2 and then proceeded with two separate interpolation techniques.

First we applied a smoothing interpolation procedure which was based on averaging the target function at every vertice of the new element of the new mesh K_{l+1} over the vertices of the parent element in the mesh K_l. Basically, this means that we get averaging and smoothing of the target function on the new mesh. However, such smoothing destroys the reconstructed function f_h which is undesirable in our adaptive algorithm. To avoid such smoothing, we propose to use barycentric

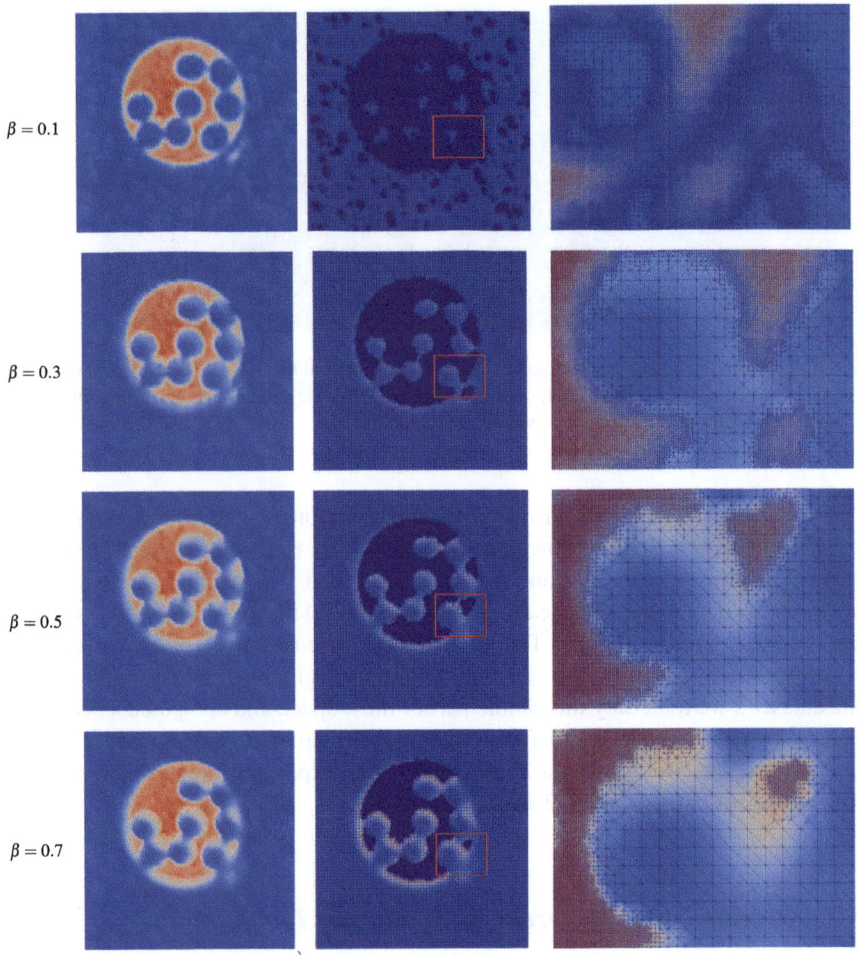

Fig. 2 Interpolation results after applying smoothing interpolation using dataset 1 on seven times refined mesh. From left to right: interpolated image without mesh, interpolated image with locally refined mesh and zoomed image with locally refined mesh. The zoomed region is denoted by a red square. From top to bottom: results with different values of parameter β in (23)

coordinates of the vertices of the new element in the new mesh K_{l+1}. The barycentric interpolation is obtained as follows

$$f(r_j) = \sum_{i=1}^{3} \lambda_i f(n_i),\ j = 1, 2, 3,$$

where $f(n_i)$ are known values of the function f at the vertices $n_i, i = 1, 2, 3$ of the parent element K in the old mesh K_l, and λ_i are barycentric coordinates of the

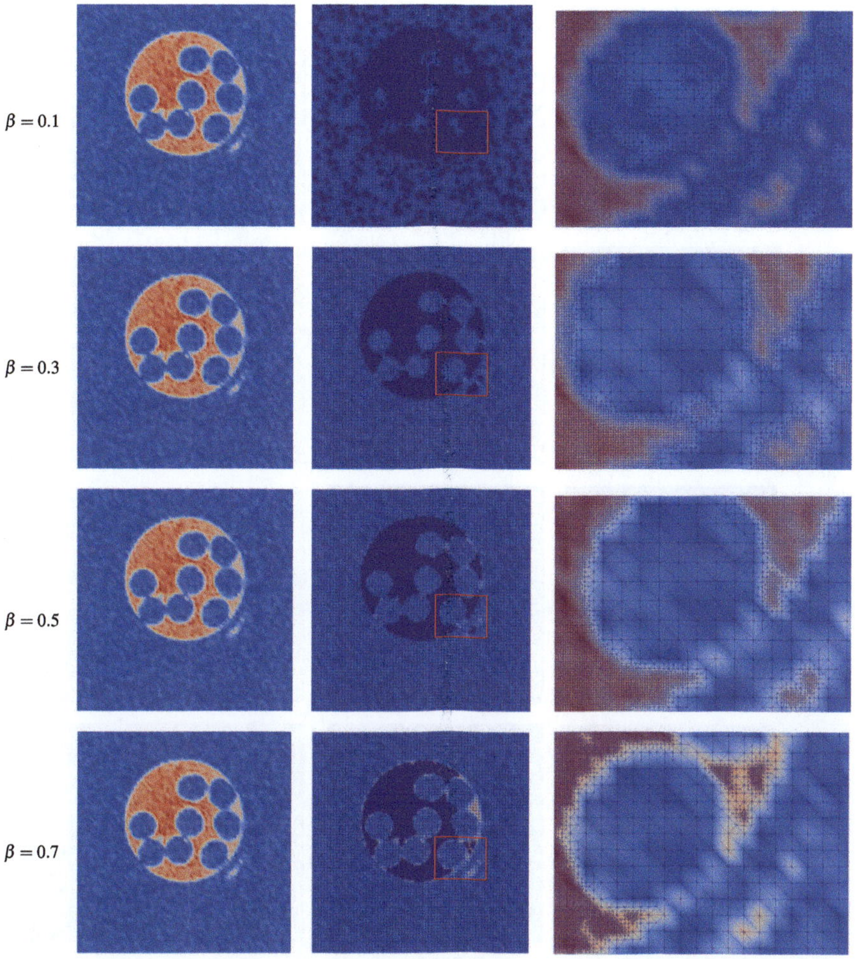

Fig. 3 Interpolation results after applying barycentric interpolation using dataset 1 on seven times refined mesh. From left to right: interpolated image without mesh, interpolated image with locally refined mesh and zoomed image with locally refined mesh. The zoomed region is denoted by a red square. From top to bottom: results with different values of parameter β in (23)

vertices r_j, $j = 1, 2, 3$ of the new element in the new mesh K_{l+1} which is contained in the parent element K of the old mesh K_l.

In our computations we refined the initial mesh seven times thus Eq. (23) of step 2 of our algorithm writes

$$|f_7| \geq \beta \max_{\Omega} |f_7|.$$

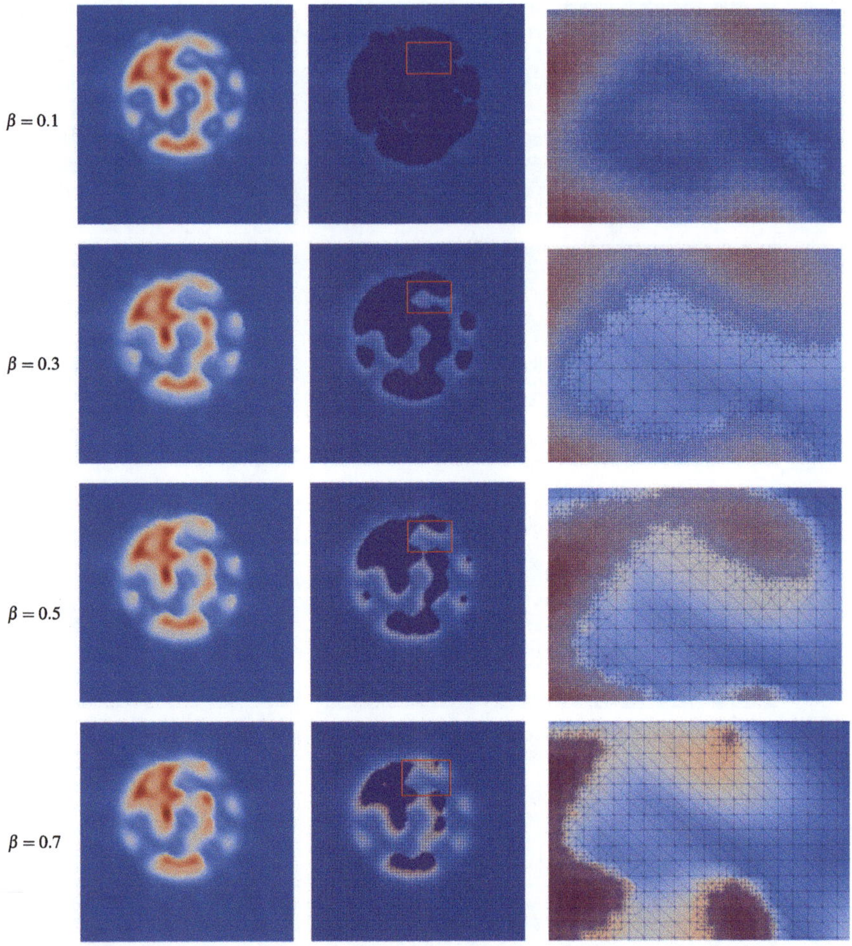

Fig. 4 Interpolation results after applying smoothing interpolation using dataset 2 on seven times refined mesh. From left to right: interpolated image without mesh, interpolated image with locally refined mesh and zoomed image with locally refined mesh. The zoomed region is denoted by a red square. From top to bottom: results with different values of parameter β in (23)

Figures 2 and 4 show results on seven times refined mesh using smooth interpolation for datasets 1 and 2. Figures 3 and 5 show results of both datasets using barycentric interpolation on seven times refined mesh. Results with both interpolation techniques are shown for several values of parameter β. The values of β are indicated on the right in Figs. 2, 3, 4 and 5.

Figures 2, 3, 4 and 5 show that the smoothing interpolation introduces blurring of the target function. Such blurring is absent when using barycentric interpolation.

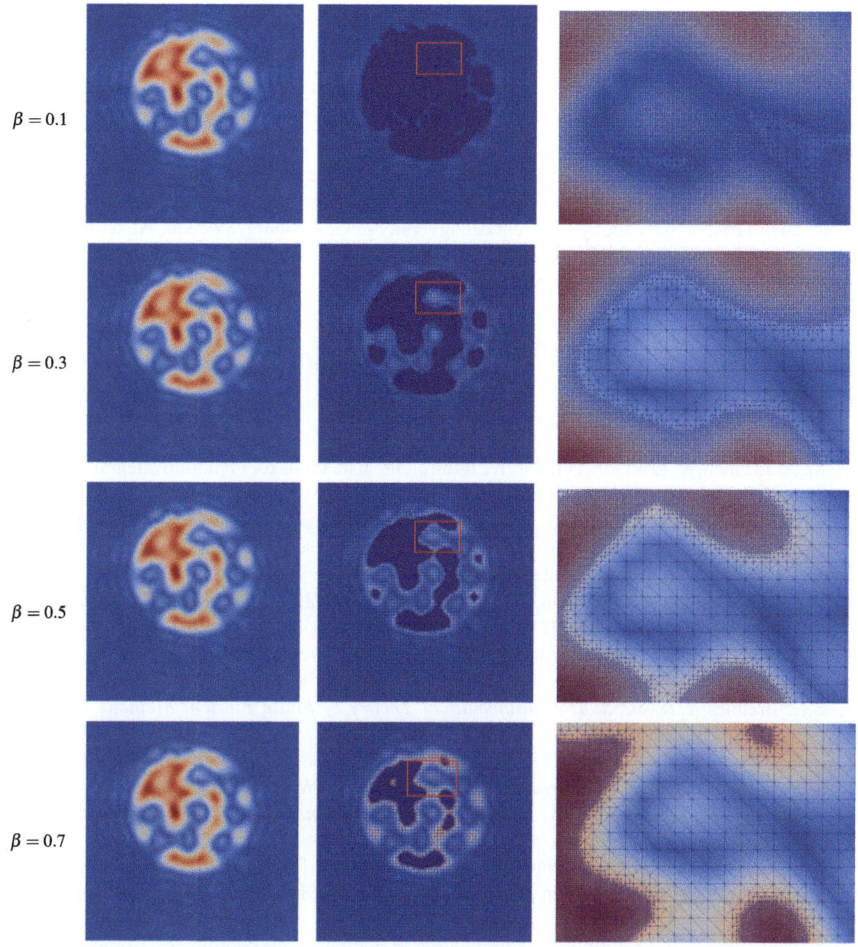

Fig. 5 Interpolation results after applying barycentric interpolation using dataset 2 on seven times refined mesh. From left to right: interpolated image without mesh, interpolated image with locally refined mesh and zoomed image with locally refined mesh. The zoomed region is denoted by a red square. From top to bottom: results with different values of parameter β in (23)

6 Conclusion

In this work we considered an ill-posed MRI inverse problem related to a Fredholm integral equation of the first kind. For the solution of this problem we formulated a finite element method and an adaptive algorithm. One of the main parts of this algorithm is interpolation of the reconstructed function on the new refined mesh. In this work we considered and compared two different interpolation techniques. First we used smoothing interpolation and then a barycentric interpolation applied to the same two data sets of MR images. The results show that the barycentric interpolation technique is preferable over the smoothing interpolation. Figures 2, 3, 4 and 5 show that the barycentric interpolation clearly improves the image quality compared to the other technique. Based on the results of this paper, in our future work, we propose to use the barycentric interpolation in reconstructing images from MR data.

Acknowledgements The research of LB was done during the sabbatical stay at the IR4M UM-R8081, CNRS, Université Paris-Sud, Université Paris-Saclay, France, which was supported by the sabbatical program at the Faculty of Science, University of Gothenburg, Sweden. We wish to thank B. Bayle and L. Meylheuc from ICube Strasbourg for providing the Osteopal samples, and L. Jourdain for her help with the NMR probe fabrication.

References

1. M. Asadzadeh, K. Eriksson, On adaptive finite element methods for Fredholm integral equations of the second kind. SIAM J. Numer. Anal. **31**(3), 831–855 (1994)
2. Yu.A. Basistov, A.V. Goncharsky, E.E. Lekht, A.M. Cherepashchuk, A.G. Yagola, Application of the regularization method for increasing of the radiotelescope resolution power, Astron. zh. **56**(2), 443–449 (1979) (in Russian)
3. A.B. Bakushinsky, M.Y. Kokurin, A. Smirnova, *Iterative Methods for Ill-posed Problems* (Walter de Gruyter GmbH& Co., 2011)
4. L. Beilina, M.V. Klibanov, Reconstruction of dielectrics from experimental data via a hybrid globally convergent/adaptive inverse algorithm. Inverse Probl. **26**, 125009 (2010)
5. S.C. Brenner, L.R. Scott, *The Mathematical Theory of Finite Element Methods*, 2nd edn. (Springer, New York, 2002)
6. R.W. Brown, Y.-C.N. Cheng, E.M. Haacke, M.R.T., R. Venkatesan, *Magnetic Resonance Imaging: Physical Principles and Sequence Design*, 2nd edn. (Wiley, 1999)
7. C. Johnson, *Numerical Solution of Partial Differential Equations by the Finite Element Method* (Dover Books on Mathematics, 2009)
8. N. Koshev, L. Beilina, An adaptive finite element method for Fredholm integral equations of the first kind and its verification on experimental data. Cent. Eur. J. Math. **11**(8), 1489–1509 (2013)
9. V. Skala, Barycentric coordinates computation in homogeneous coordinates. Comput. Graph. **32**, 120–127 (2008). Elsevier
10. A.N. Tikhonov, A.S. Leonov, A.G. Yagola, *Nonlinear ill-posed problems* (Chapman & Hall, 1998)
11. A.N. Tikhonov, A.V. Goncharsky, V.V. Stepanov, A.G. Yagola, *Numerical Methods for the Solution of Ill-Posed Problems* (Kluwer, London, 1995)
12. A.N. Tikhonov, A.V. Goncharskiy, V.V. Stepanov, I.V. Kochikov, Ill-posed problem in image processing, DAN USSR, 294(4), 832–837 (1987) (in Russian) (Moscow)

Uniqueness, Stability and Numerical Reconstruction of a Time and Space-Dependent Conductivity for an Inverse Hyperbolic Problem

L. Beilina, M. Cristofol and S. Li

Abstract This paper is devoted to the reconstruction of the time and space-dependent coefficient in an inverse hyperbolic problem in a bounded domain. Using a local Carleman estimate we prove the uniqueness and a Hölder stability in the determination of the conductivity by a single measurement on the lateral boundary. Our numerical examples show possibility of the determination of the location and the large contrast of the space-dependent function in three dimensions.

Keywords Inverse problem · Carleman estimate
Time and space-dependent coefficient · Infinite domain · Hyperbolic equation

1 Introduction

Result of this paper is based on two recent works [1, 2]. In [2] the authors deal with the inverse problem of determination the time-independent isotropic conductivity coefficient $c : \Omega \to \mathbb{R}$ appearing in the hyperbolic partial differential equation $(\partial_t^2 - \nabla \cdot \tilde{c} \nabla)u = 0$, where $\Omega := \omega \times \mathbb{R}$ is an infinite cylindrical domain whose cross section ω is a bounded open subset of $\mathbb{R}^{n-1}$, $n \geq 2$. On the other hand, in [1] the authors extend the result of [2] to a more general class of conductivities: time and space-dependent conductivities $\tilde{c}(x, t)$.

L. Beilina (✉)
Department of Mathematical Sciences,
Chalmers University of Technology and University of Gothenburg,
412 96 Gothenburg, Sweden
e-mail: larisa.beilina@chalmers.se

M. Cristofol
Aix Marseille Univ, CNRS, Centrale Marseille, I2M, Marseille, France

S. Li
Wu Wen-Tsun Key Laboratory of Mathematics, School of Mathematical Sciences,
USTC, Chinese Academy of Sciences, University of Science and Technology of China,
96 Jinzhai Road, Hefei 230026, Anhui, People's Republic of China

L. Beilina and Yu. G. Smirnov (eds.), *Nonlinear and Inverse Problems in Electromagnetics*, Springer Proceedings in Mathematics & Statistics 243,
https://doi.org/10.1007/978-3-319-94060-1_10

The goal of this work is to modify the uniqueness and stability result of [1] for the case of bounded domain Ω and a finite number of observations. Indeed, in the case of infinite cylindrical domain, some strong assumptions are required concerning the coefficient to be reconstructed (see condition (2.3) in [1]). In the current work we weaken these conditions for the case of a bounded domain which will lead to the new stability estimate. Furthermore, in a new stability inequality we have derived estimate of the distance between two sets of coefficients $\tilde{c}_1(x,t)$ and $\tilde{c}_2(x,t)$ which depends on the distance of observations at the lateral boundary of the Neumann derivative of the solutions u_1 and u_2, respectively. This stability inequality implies the uniqueness of the determination of the coefficient $\tilde{c}$. Furthermore, we can use it in numerical studies of the determination of the space and time-dependent function $\tilde{c}(x,t)$ from the backscattered noisy observations.

We consider the following initial boundary value problem

$$\begin{cases} \partial_t^2 u - \operatorname{div}(\tilde{c}(x,t)\nabla u(x,t)) = 0 \text{ in } Q := \Omega \times (0,T), \\ u(\cdot,0) = a(x),\ \partial_t u(\cdot,0) = 0 \quad \text{in } \Omega, \\ u = 0 \quad \text{on } \Sigma := \partial\Omega \times (0,T), \end{cases} \tag{1}$$

where Ω is a bounded domain and with the set of initial conditions $(a, 0)$, where $\tilde{c}$ is the unknown conductivity coefficient. We assume that $\tilde{c}$ is time and space depending conductivity such that:

$$\tilde{c}(x,t) = c_0(x,t) + c(x), \tag{2}$$

where $c_0(x,t)$ is assumed to be known. In other words, we consider the case of the perturbation of a general time and space-dependent conductivity $\tilde{c}(x,t)$ by a space-dependent one $c(x)$. In [3] the authors study a similar general non-stationary media for an inverse problem for Maxwell's equations. In [3] Lipschitz stability was also proved.

The derivation of the stability and uniqueness result of this work is based on a Carleman inequality specifically designed for hyperbolic systems, see, for example, [4–14] where have been proven stability results of the reconstruction of one or several unknown coefficients in a hyperbolic equation from a finite number of measurements. However, only theoretical investigations are presented in all these works. Theoretical uniqueness results together with numerical simulations are presented in recent works [1, 15, 16].

In numerical examples of this work the goal was to determine the conductivity function $\tilde{c}(x,t)$ that represents a sum of two space-dependent gaussians and one time-dependent function. Since by our assumption the time-dependent function $c_0(x,t)$ is known inside the domain, then we have reconstructed only the space-dependent part of the conductivity function, $c(x)$. To do that we used the Lagrangian approach together with the domain decomposition finite element/finite difference method of [17]. One of the important points of this work is that in our numerical simulations we applied one non-zero initial condition in the model problem which corresponds well to the uniqueness and stability results of this paper. Our three-dimensional examples show that we can determine the location and the large contrast of the space-dependent function. However, the location of this function in the third, x_3 direction, should be improved. This can be done using an adaptive finite element method, see details in [15, 18, 19].

The outline of the work is the following: in Sect. 2 we derive the main stability estimate and the uniqueness result, in Sect. 3 we present numerical simulations and in Sect. 4 we give conclusions to the work.

2 Main Theoretical Result

2.1 Settings and Main Result

Let $\Omega \subset \mathbb{R}^n$ be a bounded domain with smooth boundary $\partial\Omega$. We consider an acoustic equation

$$\partial_t^2 u(x,t) - \operatorname{div}(\tilde{c}(x,t)\nabla u(x,t)) = 0, \quad x \in \Omega,\ 0 < t < T. \tag{3}$$

To Eq. (3) we attach the initial and boundary conditions:

$$u(x,0) = a(x), \quad \partial_t u(x,0) = 0, \quad x \in \Omega \tag{4}$$

and

$$u(x,t) = h(x,t), \quad (x,t) \in \partial\Omega \times (0,T). \tag{5}$$

Throughout this paper, we set $\partial_j = \frac{\partial}{\partial x_j}$, $\partial_i\partial_j = \frac{\partial^2}{\partial x_i \partial x_j}$, $\partial_t^2 = \frac{\partial^2}{\partial t^2}$, $1 \le i, j \le n$.

Let $T > 0$ be given. We assume that coefficient $\tilde{c} = \tilde{c}(x,t)$ of the principal term of (3) is time and space depending in form (2) where $c_0(x,t)$ is given. In this paper, we consider an inverse problem of determination time-independent component $c(x)$ of coefficient $\tilde{c}(x,t)$, from the observations:

$$u(x,t), \quad x \in \partial\Omega \times (0,T).$$

In order to formulate our results, we need to introduce some notations. Let $x_0 \notin \overline{\Omega}$ and $\beta > 0$. We set

$$\Lambda = \left(\sup_{x\in\Omega} |x - x_0|^2 - \inf_{x\in\Omega} |x - x_0|^2 \right)^{\frac{1}{2}}. \tag{6}$$

We define admissible sets of conductivity coefficients. For arbitrarily fixed functions $\eta_0 \in C^2(\overline{\Omega} \times [0,T])$, $\eta_1 \in (C^2(\overline{\Omega} \times [0,T]))^n$ and constants $M_1 > 0, \theta_0 > 0, \theta_1 > 0$, we set

$$\mathcal{U} = \mathcal{U}_{M_1,\theta_1,\eta_0,\eta_1,c_0} = \Big\{\tilde{c}(x,t) = c_0(x,t) + c(x) \in C^2(\overline{\Omega} \times [0,T]);\ \tilde{c} = \eta_0,\ \nabla\tilde{c} = \eta_1 \text{ on } \partial\Omega \times [0,T],$$

$$\|\tilde{c}\|_{C^2(\overline{\Omega}\times[0,T])} \le M_1,\ \tilde{c}(x,t) \ge \theta_1,\ \frac{(x-x_0)\cdot\nabla\tilde{c}(x,t)}{2\tilde{c}(x,t)} < 1-\theta_0 \text{ for all } (x,t) \in \overline{\Omega}\times[0,T]\Big\}. \tag{7}$$

We note that there exists a constant $M_0 > 0$ such that $\left\|\nabla_{x,t}\tilde{c}\right\|_{C(\overline{\Omega}\times[0,T])} \le M_0$ for each $\tilde{c} \in \mathcal{U}_k$, $k = 1, 2$.

By $x_0 \notin \overline{\Omega}$, there exists a constant $\beta > 0$ such that

$$\beta + \left\{\frac{1}{2\theta_1} + \frac{1}{\sqrt{\theta_1}}\right\}\lambda M_0\sqrt{\beta} < \theta_1\theta_0, \qquad \theta_1 \inf_{x\in\Omega}|x-x_0|^2 > \beta\Lambda^2. \tag{8}$$

We point out that, if $c_0(x,t)$ satisfies $\frac{t\partial_t c_0(x,t)}{c_0(x,t)} \ge 0$ for all $(x,t) \in \overline{\Omega}\times[0,T]$, the first inequality in (8) can be replaced with

$$\beta + \frac{\lambda M_0}{\sqrt{\theta_1}}\sqrt{\beta} < \theta_1\theta_0.$$

We are ready to state our first main result.

Theorem 1 *We fix $a \in C^2(\overline{\Omega})$ satisfying*

$$(\nabla a(x)\cdot(x-x_0)) > 0 \qquad \textit{for all} \quad x \in \overline{\Omega} \tag{9}$$

Assume that $c_0 \in W^{4,\infty}(Q;\mathbb{R})$ and $\partial_t c_0(\cdot,0) = \partial_t^3 c_0(\cdot,0) = 0$ in Ω. Let $\mathcal{U}$ be defined by (7) and let $\beta > 0$ satisfy (8). We assume that

$$T > \frac{\Lambda}{\sqrt{\beta}}. \tag{10}$$

We pick arbitrarily $\tilde{c}_1, \tilde{c}_2 \in \mathcal{U}$ such that the solutions u_j, $j = 1, 2$ to (3)–(5) where $\tilde{c}_j$ is substituted to $\tilde{c}$ are in $W^{4,\infty}(\Omega\times(0,T)$. Then there exists a constant $C > 0$ such that

$$\|\tilde{c}_1 - \tilde{c}_2\|_{H^1(\Omega)} \le C\left\|\frac{\partial u_1}{\partial\nu} - \frac{\partial u_2}{\partial\nu}\right\|^{\kappa}_{H^3(0,T;L^2(\partial\Omega))},$$

where $\nu = \nu(x)$ denotes the unit outward normal vector to $\partial\Omega$ at x. Here $C > 0$ and $\kappa \in (0,1)$ are two constants depending only on Ω, T, x_0, M_1, θ_1, η_0, η_1, $a(x)$ and $\|u_j\|_{W^{4,\infty}(\Omega\times(0,T))}$, $j = 1, 2$.

The proof of Theorem 1 relies directly on the proof of Theorem 2.1, p. 4 in [1] associated with Lemma 1 which allows to work in the simpler case of a bounded domain.

2.2 The Carleman Estimate for a Hyperbolic Equation

We consider a Carleman estimate which is derived from [20].

Let us set

$$Q = \Omega \times (-T, T).$$

For $x_0 \notin \overline{\Omega}$ and $\beta > 0$, we define the functions $\psi = \psi(x, t)$ and $\varphi = \varphi(x, t)$ by

$$\psi(x, t) = |x - x_0|^2 - \beta t^2 \tag{11}$$

and

$$\varphi(x, t) = e^{\lambda\psi(x,t)} \tag{12}$$

with parameter $\lambda > 0$. We add a constant $C_0 > 0$ if necessary so that we can assume that $\psi(x, t) \geq 0$ for $(x, t) \in Q$, so that

$$\varphi(x, t) \geq 1, \quad (x, t) \in \overline{Q}.$$

Lemma 1 [20] *We assume $\mu \in C^1\left(\overline{Q}\right)$, and satisfies the following conditions*

$$\frac{\beta}{\mu}\left[1 - \frac{t\partial_t\mu}{2\mu} + \frac{|t||\nabla\mu|}{\sqrt{\mu}}\right] < 1 - \frac{(x - x_0)\cdot\nabla\mu}{2\mu} \quad for\, all \quad (x, t) \in \overline{Q}, \tag{13}$$

and

$$\mu(x, t)|x - x_0|^2 - \beta^2 t^2 > 0 \quad for\, all \quad (x, t) \in \overline{Q} \tag{14}$$

even by choosing $\beta > 0$ sufficiently small. Let $y \in H^1(Q)$ satisfy

$$\partial_t^2 y(x, t) - \mu(x, t)\Delta y = F \quad in Q \tag{15}$$

and

$$y(x, t) = 0, \quad (x, t) \in \partial\Omega \times (-T, T), \quad \partial_t^k y(x, \pm T) = 0, \quad x \in \Omega,\ k = 0, 1. \tag{16}$$

We fix $\lambda > 0$ sufficiently large. Then there exist constants $s_0 > 0$ and $C > 0$ such that

$$\int_Q (s|\nabla_{x,t} y|^2 + s^3|y|^2)e^{2s\varphi}dxdt \leq C\int_Q |F|^2 e^{2s\varphi}dxdt + Cs\int_{-T}^{T}\int_{\partial\Omega}\left|\frac{\partial y}{\partial \nu}\right|^2 e^{2s\varphi}dxdt \tag{17}$$

for all $s > s_0$.

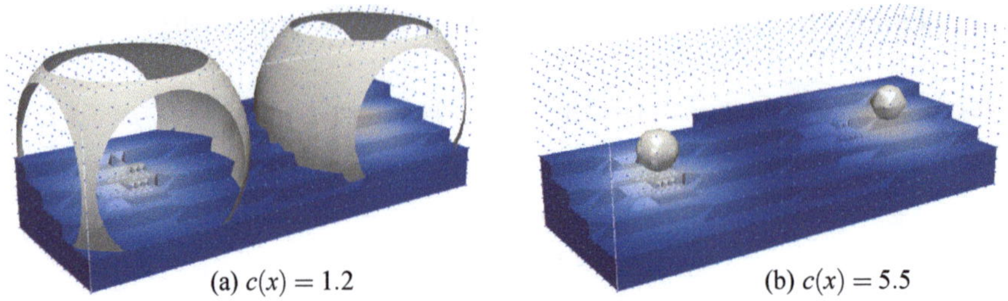

(a) $c(x) = 1.2$ (b) $c(x) = 5.5$

Fig. 1 Slices of the exact Gaussian function $c(x)$ in Ω_{FEM} given by (21)

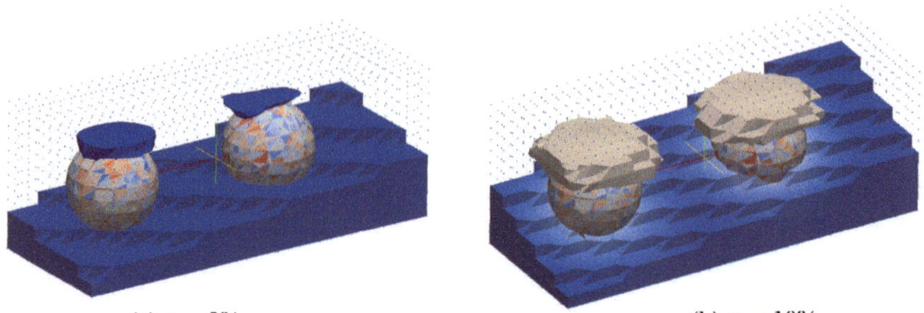

(a) $\sigma = 3\%$ (b) $\sigma = 10\%$

Fig. 2 Reconstructions obtained in Test 1 for different noise level σ in data. All figures are visualized for $c(x) = 3.5$

Proof By (13), we can see that $a_0(x, t) = 1/\sqrt{\mu(x, t)}$ satisfies (17) in P. 101 of [20]. From (14) it follows that $A(x, t; \nabla_{x,t}\psi(x, t)) = \mu(x, t)|\nabla\psi(x, t)|^2 - |\partial_t\psi(x, t)|^2$ is not zero at any point of $\overline{\Omega}$. Therefore, by Theorem 2.1 in [20] and the proof of Corollary 2.5 in [20], we can conclude that, for some sufficiently large $\lambda > 0$, there exist constants $s_0 > 0$ and $C > 0$ such that (17) holds for all $s > s_0$.

Henceforth $C > 0$ denotes generic constants which are independent of choices of $\tilde{c}_1, \tilde{c}_2 \in \mathscr{U}$ and the parameter $s > 0$ in the Carleman estimates, provided that $s > s_0$.

3 Numerical Studies

This section presents numerical examples of the reconstruction of the unknown function $c(x)$ in the Eq. (1). To do this we use the domain decomposition method of [17] and decompose Ω into two sub-regions Ω_{FEM} and Ω_{FDM} such that $\Omega = \Omega_{FEM} \cup \Omega_{FDM}$. In Ω_{FEM} we will use the finite element method (FEM) and in Ω_{FDM}—the finite difference method (FDM). We decompose also the boundary $\partial\Omega$ of the domain Ω such that $\partial\Omega = \partial_1\Omega \cup \partial_2\Omega \cup \partial_3\Omega$. Here, $\partial_1\Omega$ and $\partial_2\Omega$ are the front and back sides of Ω, respectively, and $\partial_3\Omega$ denotes the remaining part of the boundary $\partial\Omega$.

We collect time-dependent observations $S_T := \partial_1 \Omega \times (0, T)$ at the backscattering side $\partial_1 \Omega$ of Ω. We define $\Omega_T := \Omega \times (0, T)$, $S_{1,1} := \partial_1 \Omega \times (0, t_1]$, $S_{1,2} := \partial_1 \Omega \times (t_1, T)$, $S_2 := \partial_2 \Omega \times (0, T)$ and $S_3 := \partial_3 \Omega \times (0, T)$.

The model problem used in the computations is following:

$$\begin{aligned}
\frac{\partial^2 u}{\partial t^2} - \nabla \cdot (\tilde{c} \nabla u) &= 0 \text{ in } \Omega_T, \\
u(x, 0) = \theta_0(x), \quad \partial_t u(x, 0) &= 0 \text{ in } \Omega, \\
\partial_n u &= f(t) \text{ on } S_{1,1}, \\
\partial_n u &= -\partial_t u \text{ on } S_{1,2}, \\
\partial_n u &= -\partial_t u \text{ on } S_2, \\
\partial_n u &= 0 \text{ on } S_3.
\end{aligned} \tag{18}$$

In (18) the function $f(t)$ presents a plane wave which is initialized at $\partial_1 \Omega$ in time $T = [0, 3.0]$. We define it as

$$f(t) = \begin{cases} \sin(\omega t), & \text{if } t \in \left(0, \frac{2\pi}{\omega}\right), \\ 0, & \text{if } t \geq \frac{2\pi}{\omega}. \end{cases} \tag{19}$$

We initialize initial condition $\theta_0(x)$ at the backscattered side $\partial_1 \Omega$ as

$$u(x, 0) = f_0(x) = \mathrm{e}^{-(x_1^2 + x_2^2 + x_3^3)} \cdot \cos t|_{t=0} = \mathrm{e}^{-(x_1^2 + x_2^2 + x_3^3)}. \tag{20}$$

We assume that the functions $c(x) = 1$ and $c_0(x, t) = 0$ inside Ω_{FDM}. The goal of our numerical tests is to reconstruct a smooth function $c(x)$ only inside Ω_{FEM} which we define as

$$\begin{aligned}
c(x) = {} & 1.0 + 5.0 \cdot \mathrm{e}^{-((x_1 - 0.5)^2/0.2 + x_2{}^2/0.2 + x_3{}^2/0.2)} \\
& + 5.0 \cdot \mathrm{e}^{-((x_1 + 1)^2/0.2 + x_2{}^2/0.2 + x_3{}^2/0.2)}.
\end{aligned} \tag{21}$$

We also assume that the function $c_0(x, t)$ is known inside Ω_{FEM}, and we define this function as

$$c_0(x, t) = 0.01 \ \cos t \cdot \mathrm{e}^{-(x_1{}^2/0.2 + x_2{}^2/0.2 + x_3{}^2/0.2)}. \tag{22}$$

Figure 1 presents slices of the exact function $c(x)$ given by (21) for $c(x) = 1.2$ and $c(x) = 5.5$, correspondingly, and Fig. 3 presents isosurfaces of the exact function $\tilde{c}$ in the problem (18) for $\tilde{c} = 3.5$ at different times. Numerical tests of [17] show that the best reconstruction results for the space-dependent function $c(x)$ and for $c_0 = 0$ in Ω are obtained for $\omega = 40$ in (19), and we take $\omega = 40$ in (19) in all our tests.

We introduce dimensionless spatial variables $x' = x/(1m)$ and define Ω_{FEM} and Ω_{FDM} as the following dimensionless computational domains:

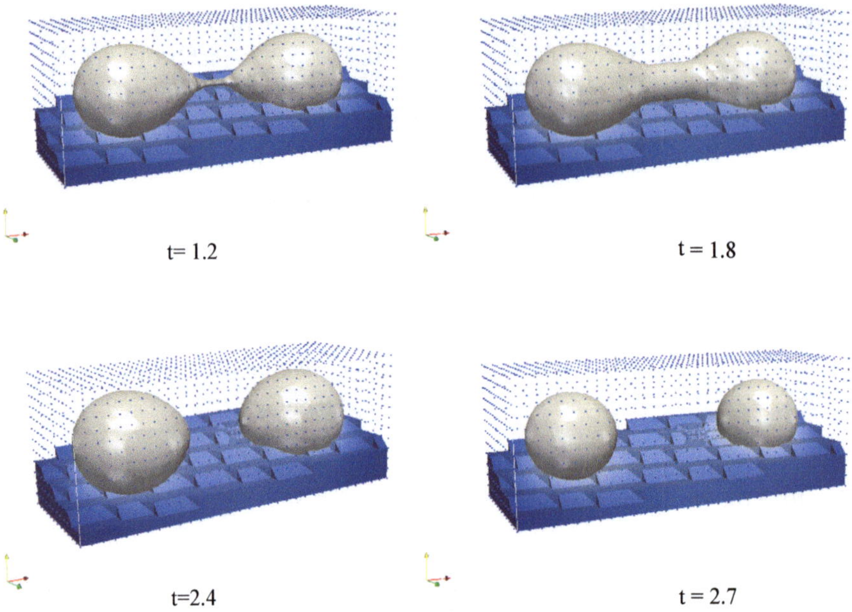

Fig. 3 Slices of the exact space and time-dependent function $\tilde{c}$ for $\tilde{c} = 3.5$ at different times

$$\Omega_{FEM} = \{x = (x_1, x_2, x_3);\ x_1 \in (-1.6, 1.6),\ x_2 \in (-0.6, 0.6),\ x_3 \in (-0.6, 0.6)\},$$
$$\Omega = \{x = (x_1, x_2, x_3);\ x_1 \in (-1.8, 1.8),\ x_2 \in (-0.8, 0.8),\ x_3 \in (-0.8, 0.8)\}.$$

We choose the mesh size $h = 0.1$ in the overlapping layers between Ω_{FEM} and Ω_{FDM} as well as in the computations of the inverse problem. However, we have generated our backscattered data using the several times locally refined mesh inside the domain Ω_{FEM}. To generate backscattered data we solve the model problem (18) in time $T = [0, 3.0]$ with the time step $\tau = 0.003$ which satisfies the CFL condition given in [21]. We also use additive noise $\sigma = 3\%, 10\%$ at $\partial_1 \Omega$. In all our computations we choose the constant regularization parameter $\gamma = 0.01$ because it gives smallest relative error in the reconstruction of the function $c(x)$. See [22–24] for other techniques of choosing regularization parameter.

We assume that the reconstructed function $c(x)$ belongs to the set of admissible parameters

$$M_c \in \{c \in C(\overline{\Omega}) | 1 \leq c(x) \leq 10\}. \tag{23}$$

To get final images of our reconstructed function $c(x)$ we use a post-processing procedure which is the same as in [17].

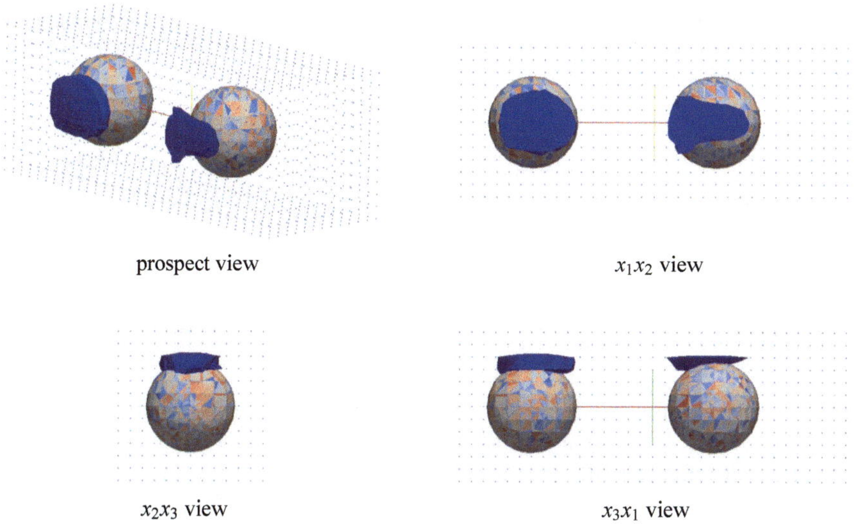

Fig. 4 Test 1. Isosurfaces of the reconstructed function $c(x)$ with $\max_{\Omega_{FEM}} c(x) = 6.66$ for $\omega = 40$ in (19). Here, the noise level in data is $\sigma = 3\%$. The spherical wireframe of the isosurface with exact value of the function (21), which corresponds to the value of the reconstructed $c = 0.7 \max_{\Omega_{FEM}} c(x)$

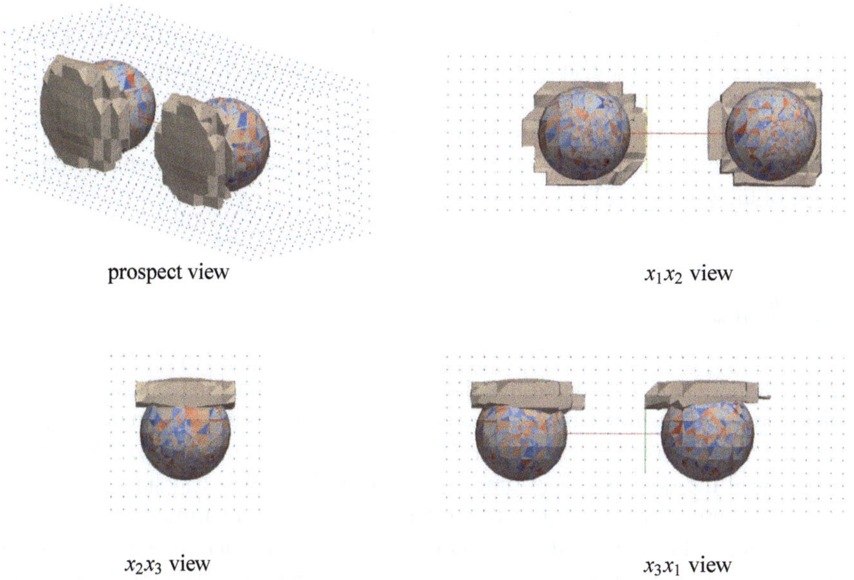

Fig. 5 Test 1. Isosurfaces of the reconstructed function $c(x)$ with $\max_{\Omega_{FEM}} c(x) = 8.11$ for $\omega = 40$ in (19). Here, the noise level in data is $\sigma = 10\%$. The spherical wireframe of the isosurface with exact value of the function (21), which corresponds to the value of the reconstructed $c = 0.7 \max_{\Omega_{FEM}} c(x)$

Table 1 Computational results of the reconstructions together with computational errors in the maximal contrast of $c(x)$ in percents. Here, $\overline{N}$ is the final number of iteration in the conjugate gradient method

$\sigma = 3\%$				$\sigma = 10\%$			
Case	$\max_{\Omega_{FEM}} c_{\overline{N}}$	error, %	$\overline{N}$	Case	$\max_{\Omega_{FEM}} c_{\overline{N}}$	error, %	$\overline{N}$
Test 1	6.66	11	13	Test 1	8.11	35	15
Test 2	4.1	32	9	Test 2	5.58	7	10

3.1 *Test 1*

In this section we present numerical results for determination the function $c(x)$ given by (21), see Fig. 1a and b, assuming, that the function $c_0(x, t) = 0$. In this case we obtain results similar to the ones given in [17].

Figures 2a, 4 display results of the reconstruction of the function $c(x)$ given by (21) with additive noise $\sigma = 3\%$ in data. Quite similar results are obtained for $\sigma = 10\%$, see Figs. 2b, 5. We observe that the location of the maximal value of the function (21) is imaged correctly. It follows from Fig. 4 and Table 1 that the imaged contrast in this function is $6.66 : 1 = \max_{\Omega_{FEM}} c_{13} : 1$, where $n := \overline{N} = 13$ is the final iteration in the conjugate gradient method. Similar observation we made using the Fig. 5 and Table 1 where the imaged contrast is $8.11 : 1 = \max_{\Omega_{FEM}} c_{15} : 1, n := \overline{N} = 15$. These figures show that the values of the background of function (21) are not reconstructed but are smoothed out since we are using the data post-processing procedure, see details in [17, 19].

Thus, we are able to reconstruct only maximal values of the function (21). Comparison of Figs. 1c and d, 4, 5 with Fig. 1a and b reveals that it is desirable to improve shape of the function (21) in x_3 direction.

3.2 *Test 2*

In this numerical test we determine the function $c(x)$ defined in (21) by using noisy backscattered data. To get reasonable reconstruction in this test we run the conjugate gradient algorithm in time $T = [0, 1.5]$ with the time step $\tau = 0.003$. We note, that we reduced the computational time compared with the first test since by running in a more longer time $T = [0, 3.0]$ we have obtained some artifacts at the middle of the domain. From other side, reducing of the computational time was resulted in obtaining of a lower contrast in the reconstructed function, see Table 1. We tested reconstruction of the function $c(x)$ with the guess values of $c(x) = 1.0$ and $c_0(x, t)$ as in (22) since by our assumption this function is known.

Figures 6 and 7 show results of the reconstruction with $\sigma = 3\%$ and $\sigma = 10\%$, respectively. We observe that the location of the maximal value of the function (21)

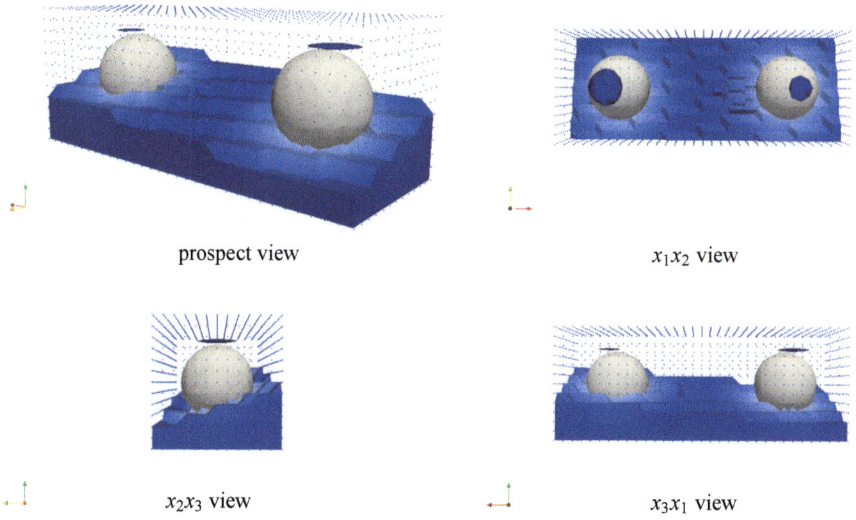

Fig. 6 Test 2. Isosurfaces of the reconstructed function $c(x)$ with $\max_{\Omega_{FEM}} c(x) = 4.1$ for $\omega = 40$ in (19). Here, the noise level in data is $\sigma = 3\%$. We outline also the spherical wireframe of the isosurface with exact value of the function (21), which corresponds to the value of the reconstructed $c = 0.7 \max_{\Omega_{FEM}} c(x)$

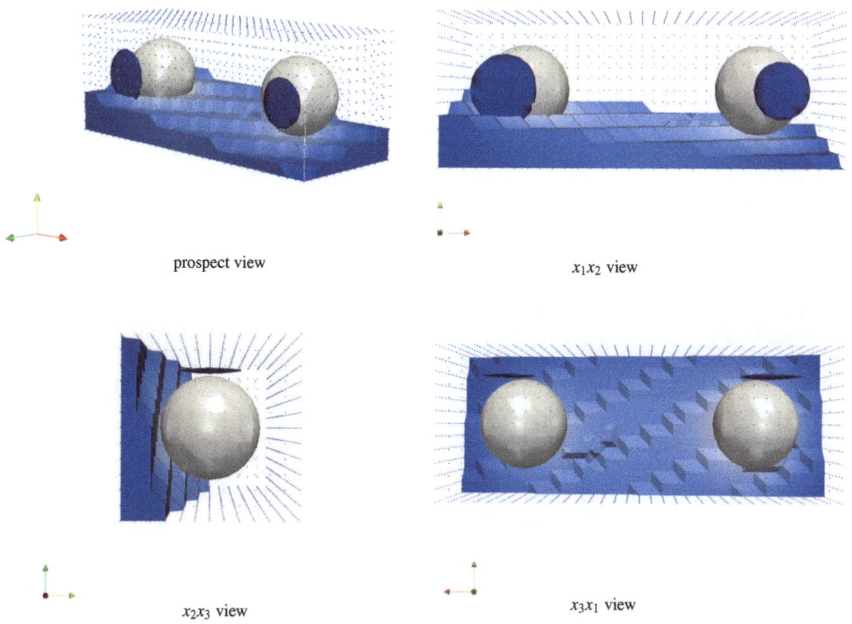

Fig. 7 Test 2. Isosurfaces of the reconstructed function $c(x)$ with $\max_{\Omega_{FEM}} c(x) = 5.58$ for $\omega = 40$ in (19). Here, the noise level in data is $\sigma = 10\%$. We outline the spherical wireframe of the isosurface with exact value of the function (21), which corresponds to the value of the reconstructed $c = 0.7 \max_{\Omega_{FEM}} c(x)$

is imaged very well. Again, as in the previous test, the values of the background in (21) are smoothed out. Comparing figures with results of reconstruction we conclude that it is desirable improve shape of the function $c(x)$ in x_3 direction.

4 Conclusions

In this work we presented uniqueness and stability results together with numerical studies of the reconstruction of the time and space-dependent coefficient in an inverse hyperbolic problem in a bounded domain. Using results of previous works [1, 2] a local Carleman estimate was proved. We note that the observation boundary used in the theorem of this work can be reduced on a part of the boundary by a special choice of the weight function, see Theorem 4.2 in [25]. In this case, one can prove the Lipschitz stability similarly with Chap. 5, Sect. 6 of [25].

Our numerical simulations show possibility of the reconstruction of function $c(x)$ in a hyperbolic problem (18) using a hybrid finite element/difference method of [17]. In our numerical tests, we have obtained stable reconstruction of the location and contrast of the function $c(x)$ in x_1x_2-directions for noisy backscattered data. The size and shape on x_3 direction should still be improved, and this can be done using, for example, an adaptive finite element method, see details in [15, 18, 19].

Acknowledgments The part of the research was done during the sabbatical stay of LB at the Institut de Mathématiques de Marseille, Aix-Marseille University, France, which was supported by the sabbatical programme at the Faculty of Science, University of Gothenburg, Sweden.

References

1. L. Beilina, M. Cristofol, S. Li, Determining the conductivity for a non-autonomous hyperbolic operator in a cylindrical domain. Math. Meth. Appl. Sci., 1–19 (2018). https://doi.org/10.1002/mma.4728
2. M. Cristofol, S. Li, E. Soccorsi, Determining the waveguide conductivity in a hyperbolic equation from a single measurement on the lateral boundary. J. Math. Control Relat. Fields **6**(3) (2016)
3. S. Li, M. Yamamoto, An inverse problem for Maxwell's equations in isotropic and non-stationary media. Appl. Anal. **92**(11), 2335–2356 (2013)
4. M. Bellassoued, Global logarithmic stability in inverse hyperbolic problem by arbitrary boundary observation. Inverse Probl. **20**, 1033–1052 (2004)
5. M. Bellassoued, Uniqueness and stability in determining the speed of propagation of second-order hyperbolic equation with variable coefficients. Appl. Anal. **83**, 983–1014 (2004)
6. M. Bellassoued, D. Jellali, M. Yamamoto, Lipschitz stability in in an inverse problem for a hyperbolic equation with a finite set of boundary data. Appl. Anal. **87**, 1105–1119 (2008)
7. M. Bellassoued, M. Yamamoto, Logarithmic stability in determination of a coefficient in an acoustic equation by arbitrary boundary observation. J. Math. Pures Appl. **85**, 193–224 (2006)
8. M. Bellassoued, M. Yamamoto, Determination of a coefficient in the wave equation with a single measurement. Appl. Anal. **87**, 901–920 (2008)

9. O. Imanuvilov, M. Yamamoto, Global Lipschitz stability in an inverse hyperbolic problem by interior observations. Inverse Probl. **17**, 717–728 (2001)
10. O. Imanuvilov, M. Yamamoto, Determination of a coefficient in an acoustic equation with single measurement. Inverse Probl. **19**, 157–171 (2003)
11. M.V. Klibanov, Inverse problems and Carleman estimates. Inverse Probl. **8**, 575–596 (1992)
12. M.V. Klibanov, Carleman estimates for global uniqueness, stability and numerical methods for coefficient inverse problems. J. Inverse Ill-Posed Probl. **21**, 477–560 (2013)
13. M.V. Klibanov, M. Yamamoto, Lipschitz stability of an inverse problem for an accoustic equation. Appl. Anal. **85**, 515–538 (2006)
14. M. Yamamoto, Uniqueness and stability in multidimensional hyperbolic inverse problems. J. Math. Pures Appl. **78**, 65–98 (1999)
15. L. Beilina, M. Cristofol, S. Li, M. Yamamoto, Lipschitz stability for an inverse hyperbolic problem of determining two coefficients by a finite number of observations. Inverse Probl. **34**, 015001 (2018)
16. O. Hussein, D. Lesnic, M. Yamamoto, Reconstruction of space-dependent potential and/or damping coefficients in the wave equation. Comput. Math. Appl. **74**, 14351454 (2017)
17. L. Beilina, Domain decomposition finite element/finite difference method for the conductivity reconstruction in a hyperbolic equation, in *Communications in Nonlinear Science and Numerical Simulation* (Elsevier, 2016). https://doi.org/10.1016/j.cnsns.2016.01.016
18. L. Beilina, Adaptive hybrid FEM/FDM methods for inverse scattering problems. Inverse Probl. Inf. Technol. **1**(3), 73–116 (2002)
19. L. Beilina, N.T. Thành, M.V. Klibanov, J. Bondestam-Malmberg, Reconstruction of shapes and refractive indices from blind backscattering experimental data using the adaptivity. Inverse Probl. **30**, 105007 (2014)
20. V. Isakov, Carleman type estimates and their applications, in *New Analytic and Geometric Methods in Inverse Problems*, ed. by K. Bingham, Y. Kurylev, E. Somersalo (Springer, Berlin, 2004), pp. 93–125
21. R. Courant, K. Friedrichs, H. Lewy, On the partial differential equations of mathematical physics. IBM J. Res. Dev. **11**(2), 215–234 (1967)
22. A. Bakushinsky, M.Y. Kokurin, A. Smirnova, *Iterative Methods for Ill-posed Problems*. Inverse and Ill-Posed Problems Series, vol. 54 (De Gruyter, 2011)
23. H.W. Engl, M. Hanke, A. Neubauer, *Regularization of Inverse Problems* (Kluwer Academic Publishers, Boston, 2000)
24. A.N. Tikhonov, A.V. Goncharsky, V.V. Stepanov, A.G. Yagola, *Numerical Methods for the Solution of Ill-Posed Problems* (Kluwer, London, 1995)
25. M. Bellassoued, M. Yamamoto, *Carleman Estimates and Applications to Inverse Problems for Hyperbolic Systems* (Springer, Tokyo, 2017)

MIX
Papier aus verantwortungsvollen Quellen
Paper from responsible sources
FSC® C105338
FSC
www.fsc.org

If you have any concerns about our products,
you can contact us on
ProductSafety@springernature.com

In case Publisher is established outside the EU,
the EU authorized representative is:
Springer Nature Customer Service Center GmbH
Europaplatz 3, 69115 Heidelberg, Germany

Printed by Libri Plureos GmbH
in Hamburg, Germany